An Introduction to Energy

Sources, Uses and Impact

Hassan Rasheed

ISBN

978-1-387-93270-2

Published in the United States of America

By Lulu.com

Table of Contents

An Introduction to Energy

Introduction

What is it all about?

Life is energy and life on Earth comes mostly from energy the Sun throws its way. Through photosynthesis plants capture this energy and thus starting its journey of flowing through us all.

Matter is in various states of potential energy. The word "potential" means that an atom or molecule can give up energy under the right circumstances. To visualize this you can think of a mountain with boulders laying in gullies on its sloping sides. Boulders at the top of the mountain have the potential of rolling down longer distances and with more energy than others that are, for example, half way up.

Some gullies are deep and hold the boulders more securely in place than others. When a boulder rolls down the mountain it moves from a higher state of energy to a lower one. One way this can happen is by applying some energy to the boulder by digging it out of its gully with a shovel. This is called applying some Kinetic energy (digging). Another way this can happen is if it rains and the side of the gully dissolves (chemically or structurally) allowing the boulder to break free and roll downhill.

In biological systems, energy is released by enzymes or other types of molecules (they are like the shovel). The energy released by each reaction is extremely small and under the control of the organism. You might say that in biological systems we talk about grains of sand instead of boulders on the side of the mountain.

An Introduction to Energy

So how do the grains of sand get up the mountain in the first place?

With the help of energy packets called photons that come from the sun and a few minute organs found in the leaves of plants carbon dioxide and water are converted to sugars which are chemical compounds with higher potential energy and therefore higher up the mountain.

The simplest of sugars is Glucose. It is a six sided ring of mostly carbon with hydrogen and oxygen hanging off the corners. From it two other compounds are made that are higher up the mountain. They are starches and cellulose.

Graph showing one way to display the composition and relationships of atoms in a sugar molecule

Starches are long chains of glucose rings and are used to store energy in the seeds of plants and other organs such as roots.

Cellulose has a bit more complicated a structure than starches and are made out of two long chains of glucose rings that are bonded together. Cellulose is used by plants to provide support structures such as stalks, branches and trunks.

Another compound that is made by the living is a fatty acid. Fatty acids are simple carbon and hydrogen strings that end in what is called a hydroxyl group of atoms. Fatty acids are used to make fat which is another compound for storing energy. Fat consists of three side by side long strings of fatty acids that are held together at one end by a group of atoms called the glycerol group.

Graph showing the composition and relationship of atoms in fatty acid molecules

___Photosynthesis

Photosynthesis is the process by which plants, some bacteria, and a few other creatures use the energy from sunlight to produce sugar, which in turn is used by plants and most other living things. This is done in plants by a green pigment called chlorophyll. Most of the time, the photosynthetic process uses water and releases oxygen. More precisely six molecules of water plus six molecules of carbon dioxide produce one molecule of sugar plus six molecules of oxygen.

___Respiration

Respiration is the process of converting the sugars produced by plants into Adenosine tri-phosphate (the universal energy chemical used by all living matter for metabolism.)

Respiration is the metabolic (building) reaction and processes that takes place in a cell to obtain energy from sugars. Energy is released by the oxidation of fuel molecules and is stored by energy carriers. The energy released in respiration is used to synthesize adenosine tri-phosphate (ATP) and its stored chemical energy can be used for many processes requiring energy, including biosynthesis, locomotion or transportation of molecules across cell membranes. ATP is also known as the "universal energy currency".

In summary, respiration takes a sugar molecule plus six oxygen molecules that produces six carbon dioxide, six-water, and a bunch of ATP molecules.

So far, we have described how the living, with the help of the sun, push matter up the mountain to store energy. The path up is not exactly the same path going down where the living start using this stored energy and create compounds like carbon dioxide, nitrous oxide and methane.

An Introduction to Energy

Before we move on there are a couple of important points we need to make. Carbon dioxide, methane and nitrous oxide exist in nature and it has devised a way to make sure that they become part of what are known as natural cycles. A cycle is a term used to describe the life of a compound from birth to death and then rebirth in a way that results in no net accumulation over the long run.

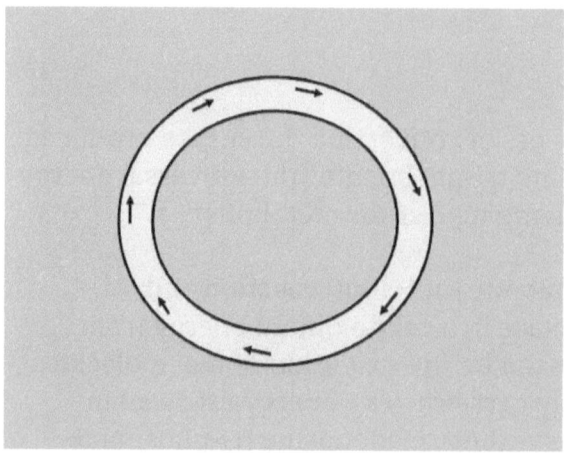

Graph showing an example of a free flowing natural cycle

Since the human population explosion occurred with its technological advances and sudden high rates of resource consumption, geologically speaking, a net increase in compounds related to the natural cycles have been produced far outpacing Earths ability to process them in the way it has for millions of years.

Definitions

With the rising cost of fuel oil in the US and the desire of us all to keep the same standard of living with comfortable homes, cars and appliances something has to give. Some people say we have to live with less but this book contends that we can keep our standard of living by being more efficient with the use of the energy that we can afford.

Top and fore most we need a reliable way to measure efficiency. The convenience of money has made it a standard unit of measure into which the value all kinds of products and services are converted, or in a more precise term "equated." Unfortunately, the process of equating everything to money is relativistic. For example, the Indian who sold Manhattan to the pilgrim from England for $15 in beads probably valued them as priceless and the pilgrim valued Manhattan in more or less similar terms.

This makes currency a poor way to measure efficiency.

Example: If you ask an owner of a 2005 Chevy Blazer how far he can travel using $2.50 in gas, he will probably say 22 miles. An Egyptian who owns the same SUV and the same amount of money would answer about 189 miles. Since we are testing the same vehicle, it is difficult to determine how a six-fold increase in efficiency materialized.

Thus it is hard to pinpoint the rock bottom efficiency of anything using currency and scientific methods are much more accurate

An Introduction to Energy

This book uses the calorie as a unit to measure energy because it is a common term used on almost all products at the grocery store. But you may ask: What is a calorie, exactly?

Definition 1:

A calorie is a unit of heat required to raise the temperature of one gram of water by one degree Celsius from a standard initial temperature

Because calories are such small quantities of energy especially when we talk about energy usage by an auto or a nation like the United States of America the following are some equations that help us grasp the measurements in better terms.

Measure	Conversion Factor	Measure
1 LBS Crude Oil	4,787,939	calories
1 Gallon Crude Oil	41,463,553	calories
1 ton oil	10,000,000,000	calories
1 ton CO2	2,804,443,487	calories

You might be confused by how Carbon Dioxide produces energy since it is the end product of combustion in things like cars and power plants. What the last figure in the table above refers to is the amount of oil that is needed to produce 1 ton of CO2. Burning that amount of oil to gain its energy also produces one ton of CO2 which is an alternate measure of oil consumption used in this book.

The other term used in this book is efficiency. Efficiency can be defined in many different ways but I will use the relationship between what we put into a process and what we get out of it. For example, if I put a rod of steel that has been heated up with one hundred calories into a glass holding fifty grams of water my expectation according to the definition above is that the water temperature would rise two degree from 50 degrees Celsius to 52 degrees Celsius. To express this transformation, I divide the result or output of the experiment by the input which is 100 calories (50 grams of water times two degrees Celsius each) by 100 calories (released by the steel rod) giving and efficiency of 1.

I should add here that I often express efficiency by the term "percent." Percentages are a common way of expressing efficiency and all we do is multiply the efficiency figure by 100. So for the above experiment an efficiency of 1 is expressed as 100%.

Definition 2:

Efficiency is measured as output divided by input multiplied by 100.

Example: If your chicken laid ten eggs but only two hatched, its efficiency is 20%.

Often times efficiency is the result of may processes in a row. In this case the total efficiency of a group of consecutive processes is the product of the efficiencies of each process. For example, if the first process is 50% efficient and the second process is also 50% then the total efficiency of both combines is 25% (50% times 50%)

Definition 3:

An increase is the production of Carbon Dioxide is an increase in energy consumption.

Often in this book estimates of fuel consumption are given in pounds or ounces of Carbon Dioxide emissions. This is another way to measure efficiency where an increase in emissions is an increase in fuel consumption.

Combustion of oil

The source of the energy that runs our economy is fossil fuels which are hydrocarbons. As the name implies, hydrocarbons consist of atoms of carbon and hydrogen in various compounds. In the process of combustion of these hydrocarbons oxygen must be added to release the energy. This reaction produces Carbon Dioxide and water.

Sources

___Crude Oil

_____The controversies

Crude oil is so pervasive in our lives and when there are any changes to its supply, price or availability we all are affected because it is the lifeblood of our economy and we tend to want to blame someone when things go "wrong". The best way to tackle the controversies is to define the four important segments of which it is made. They are:

The free market global economy

The producers

The consumers

The middlemen

There are two players in a free market economy. There is the producer who provides products and there is the consumer who needs and pays for these products. In the middle is the cost of the product. If the product is rare the assumption is that the producer is not able to produce many of them and so consumers fight over who gets to the limited supply and the one who is willing to give up more by paying more usually gets it.

If the cost of the product is too high then there is a possibility that consumers will switch from that product and chose to spend their money on something similar but less expensive. The producer may find themselves with a product that he or she can't sell. In addition, a competitor may decide to get into the production of that same product and take away sales.

If the producer is smart they will find a way to lower the price lest too many consumers turn away and switch or competitors move into his or her territory. Lowering the price may be done by lowering the cost of production through technology or improvements in efficiency. Another technique to lower the price is to sell for less by one dollar, for example, and sell larger volumes in the hope that economies of scale lower the cost of production to increase revenue.

Thus, the theory of a free-market economy promotes competition which in turn leads to efficient production techniques and lower prices.

Controversy 1: Blaming the producers

So, who are the producers of crude oil?

Only about ten percent of the countries of the world produce oil in any great quantities as can be seen in the following table.

Rank	Producing Nation	(million barrels per day)
1	Saudi Arabia (OPEC)	10.719
2	Russia	9.668
3	United States	8.367
4	Iran (OPEC)	4.146
5	China	3.836
6	Mexico	3.706
7	Canada	3.289
8	United Arab Emirates (OPEC)	2.938
9	Venezuela (OPEC)	2.803
10	Norway	2.785
11	Kuwait (OPEC)	2.674
12	Nigeria (OPEC)	2.443
13	Brazil	2.163
14	Algeria (OPEC)	2.122
15	Iraq (OPEC)	2.008

Table showing the major crude oil producing nations in 2006

The rest of the countries of the world either don't produce any crude oil or produce modest quantities that cover all or part of their needs.

Some major producing countries aren't able to meet their internal demands and so they become importers of crude oil. The United States and China are two countries that are major producers but don't meet their internal demands and so they need to import it.

Saudi Arabia is a country that produces more than it needs and so it is an exporter crude oil to countries that need it. Unfortunately, the rules of free global trade are not followed because there are political and natural resource preservation rules that come into play.

Rank	Exporting Nation	(million barrels per day)
1	Saudi Arabia (OPEC)	8.651
2	Russia	6.565
3	Norway	2.542
4	Iran (OPEC)	2.519
5	United Arab Emirates (OPEC)	2.515
6	Venezuela (OPEC)	2.203
7	Kuwait (OPEC)	2.150
8	Nigeria (OPEC)	2.146
9	Algeria (OPEC)	1.847
10	Mexico	1.676
11	Libya (OPEC)	1.525
12	Iraq (OPEC)	1.438
13	Angola (OPEC)	1.363
14	Kazakhstan	1.114
15	Canada	1.071

Table showing the leading crude oil exporting countries in 2006

For example, in 1973, some cured oil exporting countries decided to stop exporting crude oil to western countries for political reasons causing long lines at the pumps in the US.

In addition, many of the crude oil exporting countries formed a union called The Organization of Petroleum Exporting Countries (OPEC) that tries to manage the production levels and thus extract the best price for their precious non-renewable natural resource.

Unable to switch from crude oil as a source of liquid energy, people in western countries tend to blame the OPEC countries for driving up prices, in turn inflation.

Rank	Consuming Nation	(million barrels per day)
1	United States	20.588
2	China	7.274
3	Japan	5.222
4	Russia	3.103
5	Germany	2.630
6	India	2.534
7	Canada	2.218
8	Brazil	2.183
9	South Korea	2.157
10	Saudi Arabia (OPEC)	2.068
11	Mexico	2.030
12	France	1.972
13	United Kingdom	1.816
14	Italy	1.709
15	Iran (OPEC)	1.627

Table showing the major consumer countries in 2006

Controversy 2: Blaming the Messenger

The second link to the delivery chain of crude oil is the delivery. The companies that actually do the exploration for, pumping of and delivery to the consumers of crude oil are business corporations that have the goal of maximizing profits. The profit of a company is calculated as the price of the product minus cost of producing and delivery. Unfortunately, despite the fact that their profits have been around 10% and in line with many other businesses they are perceived as quasi-monopolies on account there are so few of them doing business in a multi trillion dollar business and consumers are unable to switch to any other source of fuel.

Controversy 3: The environment

There are three issues concerning crude oil and its products that effect the environment. The first is air pollution in and around our large cities in the US mainly from auto exhaust. These pollutants come from total and partial combustion of fuels. They include carbon monoxide and carbon exhaust particulates.

The second issue is the spilling of crude oil and it's by products due to human activity. About 11% of these spills occur in the drilling, refining and transportation of these products. 89% are cause by the end user in the dumping of petroleum-based products down the drains in our homes and businesses, leaky engines of autos and mishandling fuels in activities such as mowing the lawn or faulty storage tanks that leak into the ground, for example, home heating oil.

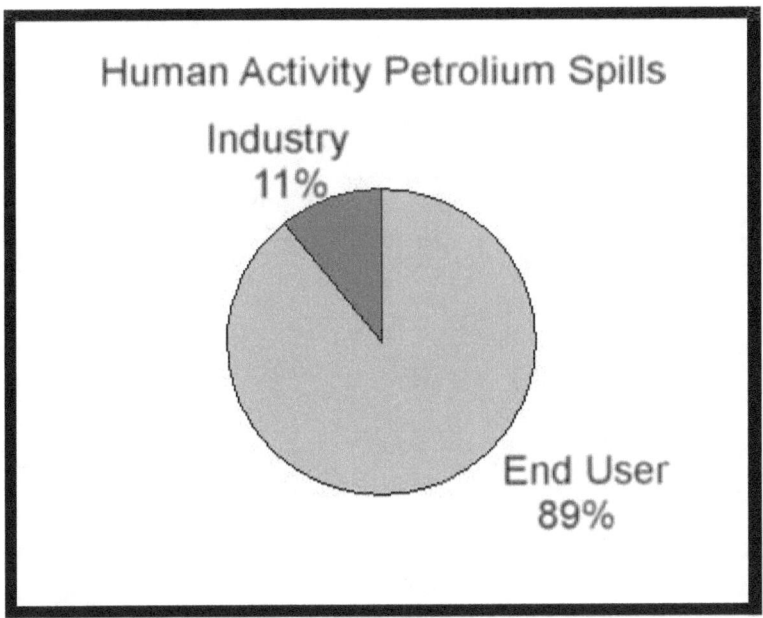

Graph showing percentages of spills due to petroleum industry and end user activities

The third issue is global warming which is mainly cause by the emissions of carbon dioxide (CO_2) from the internal combustion engine, oil fired electric generating plants and home oil heaters. Global warming is linked to the melting of the polar ice caps and the disappearance of 50% of all glaciers from the US and threatens to increase ocean levels and devastate coastal areas.

Global warming will be discussed in greater detail later in this book.

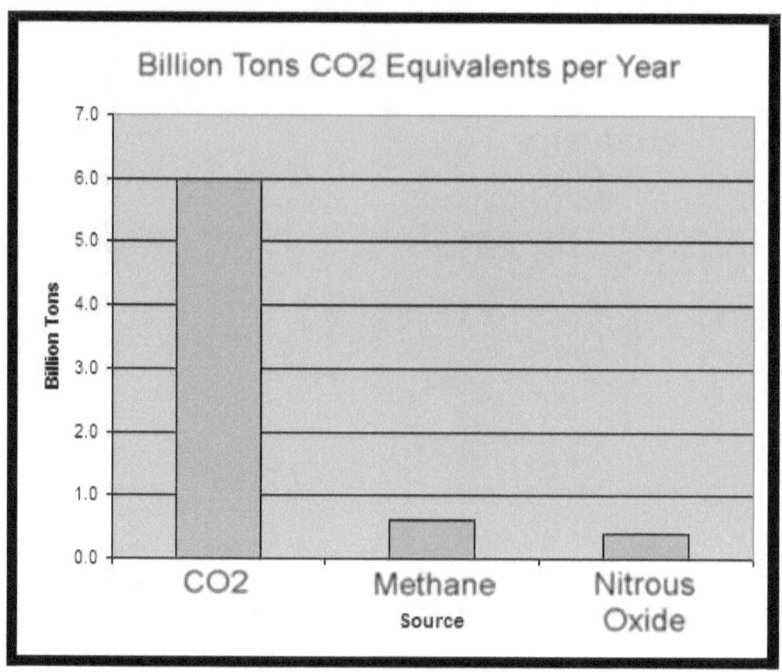

Graph showing the production of different Green House Gasses

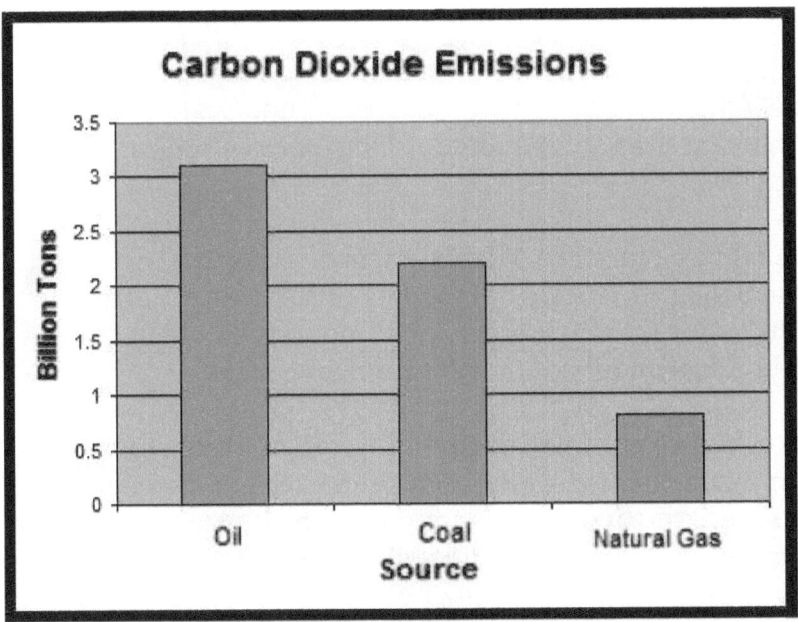

Graph showing the contributors to Carbon Dioxide emissions

In the following chapters in this section we will talk about what crude oil is, how it is made, how do we get to it and what do we do with it.

_____Crude Oil Chemistry

Crude oil is made up of hydrocarbons and hydrocarbons consist of at least one carbon and four or less hydrogen atoms. The simplest of the crude oil hydrocarbons is methane which consists of one carbon and four hydrogen atoms.

Crude oil is often described as having many different lengths of hydrocarbon chains. Some chains are open ended and are called "Alkanes," others that are closed ended called "Cycloalkanes" and "Asphaltenes" are more complicated structures.

Crude oil has a mixture of hydrocarbon chains that are from one carbon atom (methane) to forty or fifty carbon atoms in length. Crude oil also has other compounds mixed in that contain nitrogen or sulfur atoms to name a few.

If you take two methane compounds and put them together by removing one hydrogen atom from each and connecting them where the hydrogen was you get the hydrocarbon chain of ethane. In a similar fashion you can create the hydrocarbon chain of propane by adding a methane atom to one end of an ethane chain.

The color of crude oil differs greatly from black to brown to gray depending on the quantity of the different hydrogen chains it contains and other compounds.

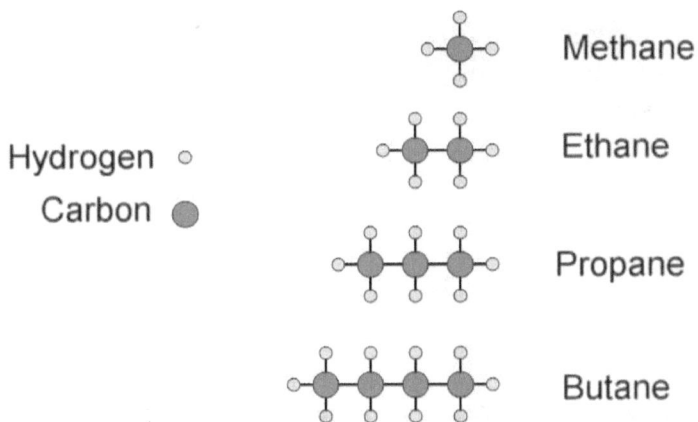

Hydrogen ○
Carbon ●

Methane

Ethane

Propane

Butane

Graph showing four aromatic hydrocarbon chains found in crude oil

The boiling point of hydrocarbon chains

Different hydrocarbon chains found in crude oil have different boiling temperatures and generally, a chain with four of less carbon atoms boils at room temperature and therefore is a gas. Hydrocarbon chains with from five to twelve carbon atoms are liquid at room temperature and much longer hydrocarbon chains are usually solid.

This characteristic of hydrocarbons is useful in separating the different length chains from each other as will become apparent in subsequent chapters.

Viscosity

Generally, the longer the hydrocarbon chain the more viscose it is so a hydrocarbon chain with twenty carbon atoms is as thick as molasses.

How Crude Oil was formed

About 300 to 400 million years ago there existed plants and animals in the oceans of the world. As they died, fell to the bottom where there were anaerobic (absence of oxygen) conditions they accumulated over millions of years. Eventually, they were buried in one way or another by sand, silt and rock.

As the layers of sand and rock accumulated to depths of 3 miles or more great pressures were brought upon these organic remains. Heat is also produced under these conditions and the combination transformed these organic remains first to a waxy organic substance that is currently found in what is known as shale rock. Further heat and pressure transformed this waxy substance into oil and then gas.

This oil is lighter in density from the surrounding rock and sand so it migrated up and when it encountered obstacles it also moves laterally for hundreds of miles. Most of it found its way to the surface of the earth where it is digested or consumed by bacteria while some if it got trapped in reservoirs under impervious rock and sits there until we discover it.

These oil reservoirs are most commonly found above layers of water at the bottom then the lighter oil. Natural gas is found above the oil.

_____Finding the Oil Reservoirs

The most convenient places to find oil is in traps consisting of some sort of dome formation of high density impenetrable rock that obstructs oil and gas migrating any further upward and so they accumulate. These traps are deep underground and it takes great skill and luck to find them.

Graphic showing where oil is trapped underground under impenetrable domes of rocks

An Introduction to Energy

Geologists are assigned the task of finding the oil. They look for the right conditions under which oil reservoirs may exist underground. In the old days, the method was to interpret surface feathers to indicate the possible existence of these oil "traps" that include soil type and shallow drilling to examine core samples. Today, a variety of devices and other techniques are employed to find these reservoirs. In addition to the old methods, satellite images are used along with gravity and magnetic meters that measure tiny movements underground of possible flowing oil. They also use so called "sniffers" that can detect minute amounts of hydrocarbons escaping from the surface of the ground.

The most commonly used method however is seismology. Seismology is the science of interpreting the reflection of shock waves from rock and liquid features underground. The equipment used is able to show the layering of deep rock formations and from those images the geologist gives his best estimate as to the existence of oil or not.

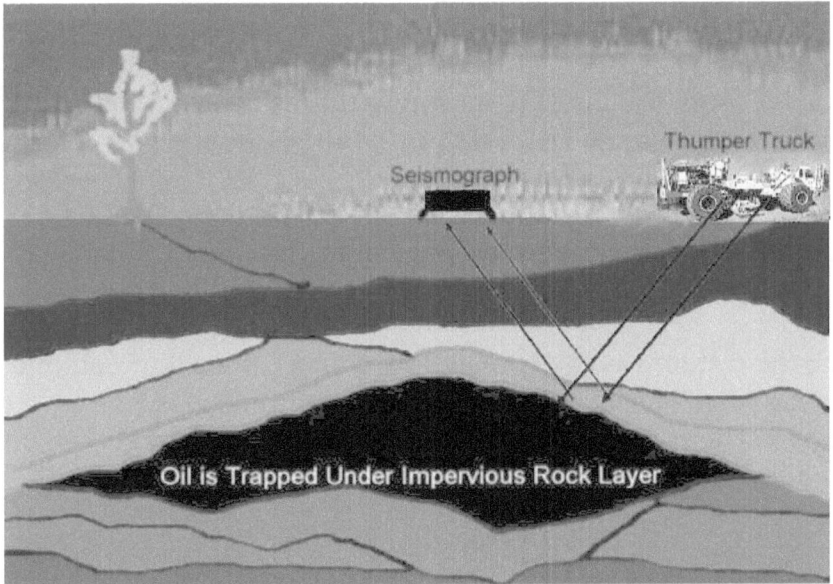

Graph showing a thumper truck and a seismograph testing to the existence of an oil reservoir

Shock waves are created by explosions or pounding on the earth with what is known as "thumper trucks" over land and compressed air guns floating over water sites like seas, oceans or lakes. Different rock formations and liquid reservoirs reflect these shock waves in different ways according to their densities and thus seismologists are able to interpret the data that is gathered as to where there might be oil. New wells are discovered about 10% of the time.

After the Thumper truck stops thumping and the geologist and seismologist gather the data there are three steps in evaluation that need to take place before they drill for oil. It may help to look at the decisions made to decide to drill for the Alaska oil in 1995 in ANWR.

An Introduction to Energy

The first step taken was number crunching of the data recovered from the seismology tests. The geologist then made a statistical guess as to where there is oil and how much. In Alaska the estimate was about 21 billion barrels.

The next step was to estimate how much of that oil can be pumped out using the technology and equipment available at that time. The estimate came back about 7.7 billion barrels.

The third step was to estimate how much can be pumped out for the price of a barrel of oil which at the time was $24 a barrel. The answer came back about 5.2 billion barrels can be pumped profitably.

Revenue	$124,800,000,000.0
Cost of drilling	-$12,480,000,000.0
Cost of Pipeline	-$24,960,000,000.0
Cost of Shipping	-$24,960,000,000.0
Cost of Operation	-$49,920,000,000.0
Profit	$12,480,000,000.0
Profit per Barrel	$2.4

Table showing an estimate of the profitability of pumping 5.2 billion barrels and selling a barrel for $24

_____Drilling for Oil

Once the site has been selected there are four things that must be done before the actual drilling for oil can begin which are legal issues:

1- A survey is done to determine its boundaries

2- Environmental impact studies may be done

3- Legal evaluation of lease agreements, titles and right-of way accesses for the land are obtained

4- Legal jurisdiction must be determined for off shore rights

Once the legal issues are out of the way the land can be prepared. First access roads need to be built to bring in the equipment. Second the land must be cleared and leveled. Third a source of water must be found because it is an essential part of the drilling operation. If it is not available then a well is dug to provide it. Fourth a waste reservoir pit is dug and lined with plastic to dispose of rock cuttings and drill mud if it is allowed otherwise this material must be trucked out.

Once the land has been prepared, a rectangular pit is dug that they call the cellar around the location where they will dig or drill the hole from which oil will be pumped. The cellar must be big enough to house the oil drilling rig and associated equipment, mud reservoir and drilling accessories. The first part of the oil hole is wide and shallow and is lined with what is known as the conductor pipe. Other holes are dug for supplementary equipment.

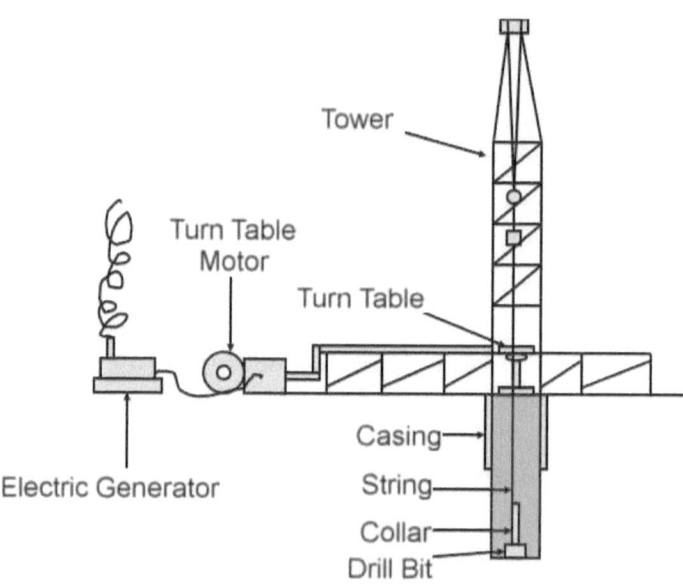

Graph showing a typical crude oil drilling rig and some of its components.

Once the cellar and the conductor pipe are in place the drilling rig is brought in by truck, barge or flown in depending on the terrain and put together. First the drill bit is fastened to the drill string which consists of a pipe that is lowered into the drill hole. The drill string is rotated using a turn table which is rotated by an electric motor. The electric motor is in turn powered by a powerful diesel electric generator. The drill bit and string are lowered through a blowout vale to contain the mud pressure systems and possible blowouts when oil is found. As the drilling continues more pipes are fastened to the drilling string using threaded collars. The pipes are hoisted up the derrick using power from the electric generator and placed on top of the last segment of the drill string.

The drill bit is a specialized piece of equipment to cut through rock and can be made of diamonds, tungsten steel or carbide steel. The oil drilling hole is continually lined with a large diameter concrete pipe called a casing to prevent the walls of the drill hole from collapsing and allows drilling mud to circulate around the drill string.

As drilling continues, a fine slurry of mud is pumped under high pressure down the string pipe and exits through the drill bit and carries the broken earth and rocks up the outside of the drill string and inside the casing. As the rock and mud emerges from the drilling hole it goes through a sieve to remove the rock and then recycled to the mud slurry hopper and used again. The rocks are collected and deposited in the plastic line waste pit or reservoir.

The drilling continues as new sections of pipe are added to the drill string and once they reach a predetermined depth close to where they think the oil reservoir cap rock is they stop. They then cement the oil-hole casings together by pumping cement around the outside of the casings and letting it harden.

Once the predetermined depth has been reached or they hit the cap rock they remove the drill and drill string from the hole and take samples of soil to confirm that they have reached the cap rock or the oil find.

The crew then completes the well by lowering a perforating gun to the production depth. The gun has explosive charges to create holes in the casing. After the casing has been perforated, they run a small-diameter production pipe into the hole as a conduit for oil and gas to flow up. Then a device called a packer is run down the outside of the tubing. When the packer is set at the production level, it is expanded to form a seal around the outside of the production pipe. Finally, they connect a structure called a Christmas tree to the top of the production pipe and cement it to the top of the casing. The Christmas tree allows them to control the flow of oil from the well.

To start the flow of oil into the well acid is pumped down and out the perforations for a limestone oil reservoir which dissolves channels in the limestone that lead oil into the well. For sandstone reservoir rock, a specially blended fluid containing sand, walnut shells, aluminum pellets is pumped down the well and out the perforations. The pressure from this fluid makes small fractures in the sandstone that allow oil to flow into the well. Once the oil is flowing, the oil rig is removed from the site and production equipment is set up to extract the oil from the well.

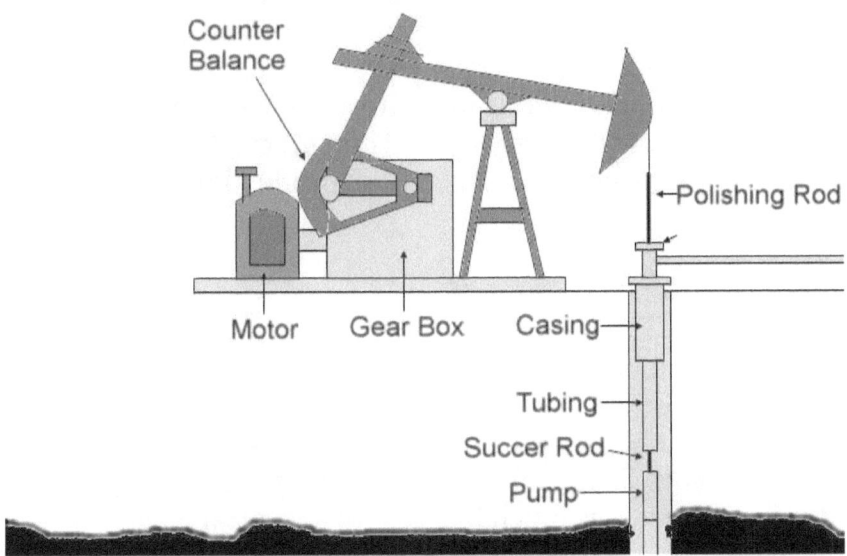

Graph showing an oil pump

The production equipment consists of an electric motor that drives a gear box that in turn moves a lever. The lever pushes and pulls a polishing rod and sucker rod up and down. The sucker rod is attached to a pump. This system forces the pump up and down, creating a suction that draws oil up through the well.

If the oil is too thick flow a second hole is drilled close by into the reservoir and steam is injected under pressure which thins the oil in the reservoir, and the pressure helps push it up the well. This process is called enhanced oil recovery.

Alternative and new drilling methods

Because of the damage to the environment and high cost of vertical drilling, new drilling methods are being developed.

The most popular new method is directional drilling. This includes all forms of drilling in which the hole is slanted, horizontal, or curved from the drilling site. Thus, oil reserves in extremely sensitive ecosystems can be reached while most of the drilling equipment is miles away including shallow off-shore deposits. An added benefit is that it can tap even very low, heavy oil reserves and increases production from a single oil well dramatically.

Types of Directional Drilling

Slant drilling is where the hole is drilled at an angle with a special drill bit. The resulting well is straight, so the extraction process is much like that of a vertical well.

Horizontal drilling is where a slant well is drilled that deviates more than 75 degrees from the initial vertical hole. This is the most common type of directional drilling.

Deviated S-turn drilling is a combination of vertical and horizontal drilling that allows greater distances between the origin and the deposit.

_____Transportation of Crude Oil

Pipelines

Oil pipelines are made 4 to 47 inch in diameter steel or plastic tubes that are generally buried 3 to 6 feet underground (depending on environmental impact statement.) Depending on the wax content of the crude oil or the temperature wax buildup may occur within a pipeline and devices called "scrapers" are employed to clear the wax from the pipes. Scrapers are launched from pig-launcher stations and travel through the pipeline to be received at any other station down-stream.

Types of Pipelines

In general, pipelines are classified in three main categories depending on their main function:

1. Gathering Pipelines – A group of smaller interconnected pipelines forming complex networks with the main purpose of bringing crude oil from several wells to a processing facility. These pipelines are usually short, couple of hundred of meters long, and have smaller diameters.

2. Transportation Pipelines - Mainly long pipes with large diameters, moving crude oil between cities, countries and even continents.

3. Distribution Pipelines – A group of several interconnected pipelines with small diameters, used to take the products to the final consumer. Basically, pipelines at terminals to distribute final products to tanks and storage facilities are included in this group.

Pipeline Components

Pipeline networks are made of different equipment that operate together to move crude oil from location to location. The main components are:

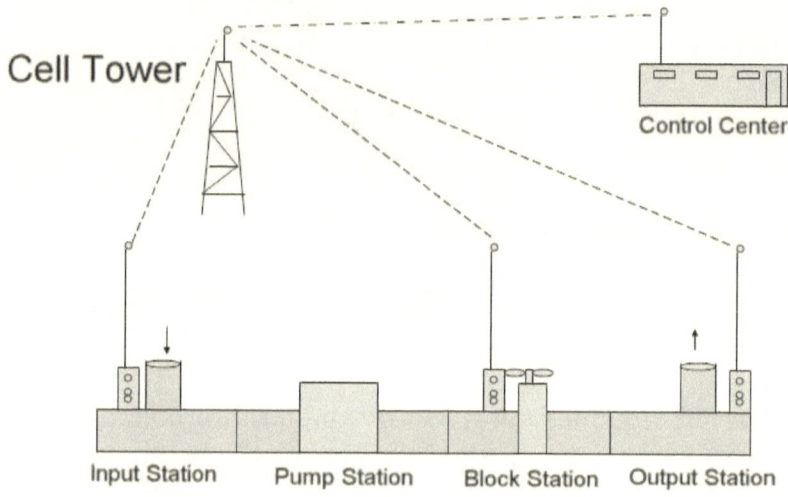

Cell Tower

Control Center

Input Station   Pump Station   Block Station   Output Station

Pipeline Schematic

- Initial Injection Station - Known also as Supply station, is where the product is injected into the line.

- Pump Stations - Pumps are located along the line to help move the product through the pipeline. The topography of the terrain, the type of product being transported, or operational conditions of the network define where these pumps are located.

- Partial Delivery Station - Known also as Intermediate Stations, these facilities allow the pipeline operator to deliver part of the product being transported.

- Block Valve Station - These are valves by which the operator can isolate any segment of the line to perform some specific maintenance work or isolate a rupture or leak. Block valve stations are usually located every 20 to 30 miles.

- Regulator Station - This is a special type of valve station, where the operator can release some of the pressure built into the line. Regulators are usually located at the downhill side of a peak.

- Final Delivery Station - Known also as Outlet stations or Terminals, this is where the product will be distributed to the final consumer.

Pipeline Regulation

In the U.S. pipelines are regulated by the Pipeline and Hazardous Materials Safety Administration and offshore pipelines are regulated by the Minerals Management Service. In Canada pipelines are regulated by the provincial regulators or by the National Energy Board if they cross borders.

Pipeline Operation

When a pipeline is built it is basically a network that requires a type of network of field devices consisting of controllers and sensors that support remote operation.

Field devices are basically data gathering instruments and communication systems. The field Instrumentation includes flow, pressure, temperature gauges/transmitters, and other devices to measure the relevant data required to operate thy system. These devices are installed along the pipeline at specific locations, such as injection or delivery stations, pump stations, and block valve stations.

The information measured by these field instruments are transmitted by satellite channels, micro wave links or cellular phone connections.

Pipelines are then controlled and operated remotely, from The Main Control Room where field measurements are consolidated in one central database.

Tankers

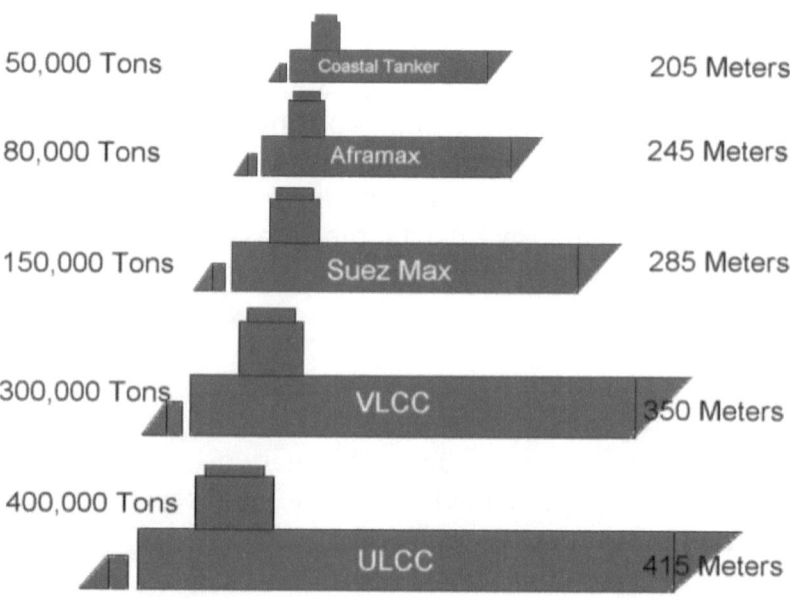

50,000 Tons	Coastal Tanker	205 Meters
80,000 Tons	Aframax	245 Meters
150,000 Tons	Suez Max	285 Meters
300,000 Tons	VLCC	350 Meters
400,000 Tons	ULCC	415 Meters

Graph showing various crude oil tanker classes and their capacities

Demand for oil was encouraged by the invention in 1897 of the Diesel engine, which used oil as a fuel rather than coal. In

1911, the first diesel powered ship crossed the Atlantic. By 1927 some 28% of the world merchant fleet used oil for power.

During the next few decades, oil replaced coal as a source of energy and tankers soon formed a major portion of the world fleet. In 1950 the standard sized was the "T2" oil tanker 620 of which were built in the United States between 1942 and 1946. Many T2 ships were sold after the end of hostilities of World War II and formed the backbone of many fleets. They had a dead weight of 16,000 tons and many were still being used in the 1960s. Tanker sizes grew and in 1959 the 114,356 dead weight Universe Apollo became the first tanker to pass the 100,000-ton figure: within a decade ships five times that size were being planned to take advantage of economies of scale.

Crude oil tankers were relatively unsophisticated and fairly simple to build. And, thanks to the square/cube rule, it pays to build them big. This rule describes the relationship between the volume and surface area of a ship. The surface is the area that is in contact with the water where friction occurs and in turn more fuel is required to move the ship. The larger the oil tanker the percentage of the surface area decreases with respect to the volume. For example, if two ships are built, one with sides 20 meters long and the other with sides 40 meters long, the surface area touching the water of the first will be 400 square meters and that of the second 800 square meters, or twice as big. But the volume of the first box will be 8000 cubic meters and that of the second 64000 cubic meters, or eight times as great.

For example, a 60,000 dead weight tanker might need about 16,000 horsepower to operate at 15 knots. A 260,000 dead weight tanker might require 42,500 horsepower thus 2.7 times the energy would enable more than 4.3 times as much cargo to be transported.

Since it is the amount of steel used that basically determines the cost of constructing the ship, it can be seen that using twice the steel will enable carrying eight times as much cargo. There are other advantages to be gained from building ships bigger.

In addition, crew costs do not rise in proportion to the size of the ship. In fact, from the 1950s onwards crew sizes steadily decreased, as owners took advantage of automation and other technical advances. By the 1980s 200,000 dead weight tankers, or more, were operating with crews of 24, compared with the 45 required to operate a T2 tanker thirty years before. Other personnel costs, such as shore management, also tended to stay the same, or to fall, since the number of people required to run a fleet depends mainly on the number of ships involved rather than their tonnage.

However, tanker size did not grow and grow. In the first place, there was a limit to the number of shipyards capable of building them and the number of ports able to receive them. Secondly, many of the world's most important shipping routes were unable to cope with very large ships. The Suez Canal, located on what was the most important shipping route in the world in the 1960s was limited to 70,000 dead weight tankers. The Malacca Strait, separating Malaysia from Indonesia, is too shallow for anything greater than 260,000 dead weight tankers. Larger tankers traveling from the Gulf to Japan, for example, have to go via the Lombok Strait, which adds and extra 1,100 miles to the voyage. Many other straits, such as the Straits of Dover and the Bosporus, present navigational difficulties to large ships.

The most important factor that encouraged very large ship building was the closure of the Suez Canal in 1967. This meant that ships going from the Gulf to Europe and North America had to go around the Cape of Good Hope instead. At the same time, business and trade were generally booming and, for the first time, the United States had become a major oil importer instead of exporter. Freight rates soared and so did profits. At one time, it was possible for the cost of a new 200,000 dead weight VLCC (Very Large Crude Carrier) to be paid off in one year.

_____Refining Crude Oil

Distillation

Distillation is the process of separating two different liquids from each other when they are in a mixture. When two different liquids at room temperature coexist they often exhibit two different boiling points when the temperature rises slowly. If you hold the temperature of the liquid at the boiling point of one it will evaporate into the air.

The process of distillation is one where the boiled off liquid or the one that evaporated is captured and subjected to a cooling process to return it to liquid form once again and separate it from the original liquid mixture.

For example, if you have a mixture of ethanol and water in a flask, you can raise the temperature to 78 degrees Celsius which is the boiling point of ethanol. Gradually the ethanol will evaporate and if you arrange it so that you capture the escaping evaporative gas in another flask that is cooled the ethanol will condensate into a liquid once again and thus you have separated the two liquids.

Fractional distillation

Fractional distillation is the process of separating more than two substances from each other using distillation. For example, of you have a mixture of oil, water and ethanol and you want to separate them then the mixture is boiled to 105 degrees Celsius and the vapors are passed through two flasks the first cooled to 98 degrees Celsius and the second cooled to 70 degrees Celsius water will be distilled in the first flask and the second flask will distill out the ethanol.

Fractional distillation is the process used to separate out the different useful products found in crude oil.

Crude oil refineries

Crude oil refineries use fractional distillation to separate out the different products of crude oil the major ones being fuels that provide the energy for our autos and heating. A refinery usually consists of one or more distillation towers and many holding tanks for the different products that are separated out.

The distillation tower is air cooled meaning that no active refrigeration technology is used to control the temperatures of the various chambers in which the different products distill out.

The bottom chamber is where heated crude oil is piped in and maintained at three hundred and fifty degrees Celsius. As the crude oil evaporates it is cooled gradually until it reaches the top of the tower at twenty five degrees Celsius.

From the top of the refinery fractional distillation tower petroleum gas is collected and piped to a refrigeration unit that converts it into a liquid that then is distributed as liquefied petroleum gas (LPG).

The next chamber down the tower is where benzene or gasoline is distilled out and collected. The next chamber down is where naphtha is distilled out. Naphtha is a compound used in the manufacture of many chemicals.

In the next chamber down is where kerosene is distilled out and delivered to airports to be used as fuel for airplanes. In the next chamber down diesel fuel is collected and distributed to be used for trucks, busses, cars and electrical power generating plants.

In the next chamber down fuel oils are distilled out (sometimes called "bunker oil") that fuels ships and electric power stations. And finally, that which remains at the bottom of the distillation tower is asphalt and roofing tar.

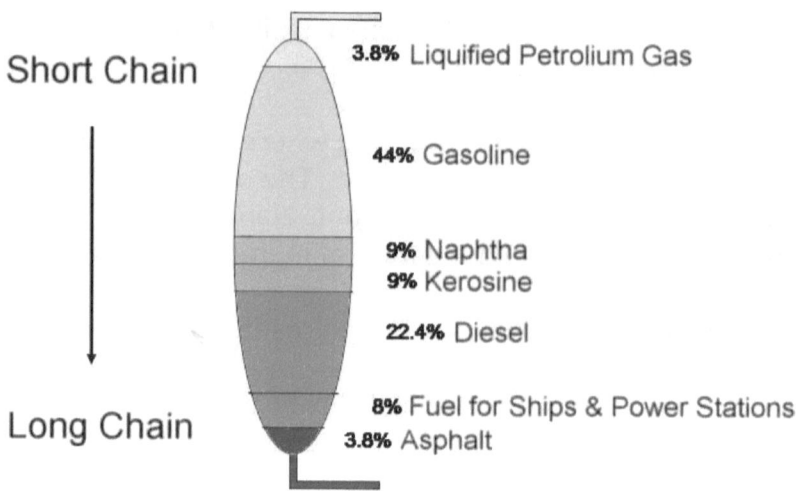

Graph showing average fractional distillation products of US crude oil and their uses

_____How Much Oil is Out There?

Oil reserves is a term used to describe the statistical probability of finding oil, drilling for it, pumping it, and delivering it to the customer at a price that customer is willing to pay at any one given time. In addition it describes the feasibility of current technology to find and pump it at current prices, with current commercial terms and government consent.

There are three main categories that the crude oil industry uses to describe these statistical probabilities. They are:

Proven Reserves - defined as oil that is "Reasonably certain" to be in existence and is usually given a probability of delivery to the customer of 90%. These reserves can be "Proven and Developed" meaning that they are ready for the market or they can be termed "Proven but Undeveloped" which are reserves requiring additional capital investment (drilling new wells, building pipelines, etc.).

Country	Proven Reserves (Billion Barrels)
Saudi Arabia	262
Canada	180
Iran	133
Iraq	112
United Arab Emirates	98
Kuwait	97
Venezuela	77
United States of America	21
Mexico	14

Table showing proven reserves of the major crude oil producing countries

Probable Reserves - defined as oil "Reasonably Probable" of meeting market demand and price and usually given a probability of delivery to the customer of around 50%. This type of reserve is usually undeveloped and required infusions of resources such as cash and technology to make it available to the consumer. In general it is harder to get to because if is found in less than ideal locations.

For example, there is a huge amount of oil in small and scattered deposits that would cost more to drill for using conventional means than the oil is worth on the market today. In these cases directional drilling may be employed to tap into them. Directional drilling is where a single drilling rig is erected and from it several lateral wells are drilled in the direction of the scattered deposits.

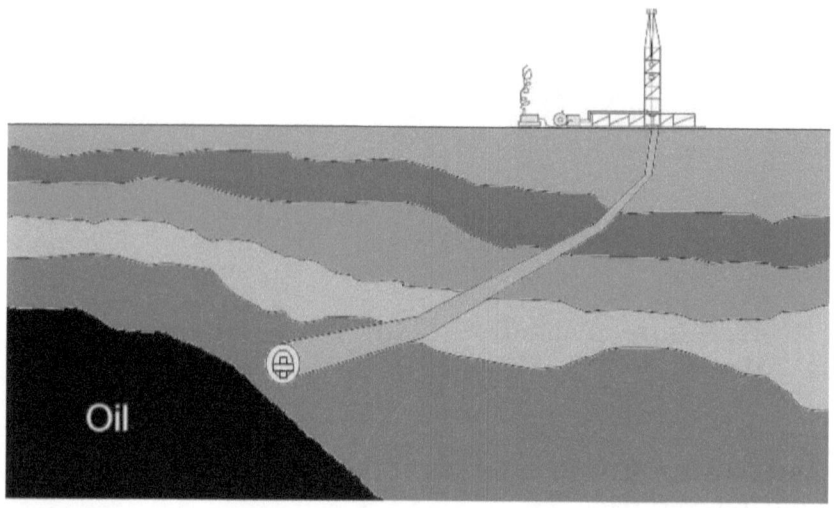

Graph showing directional drilling

Another type of oil exists that is thick in consistency, in sandy soils and hard to pump out. In this case a method called "Steam Flooding" is employed in which steam is injected in an adjacent well that softens the oil and makes it fluid to flow up the production well.

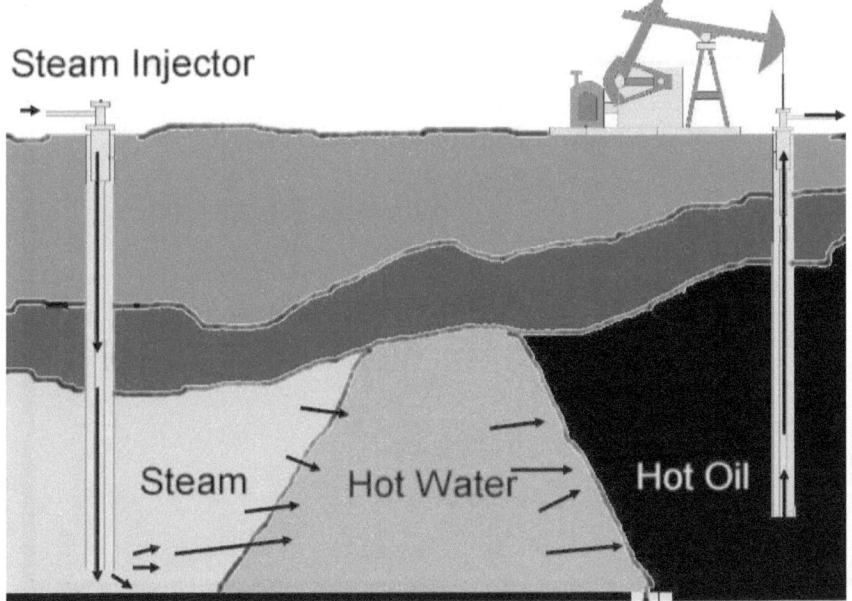

Steam Injector

Steam   Hot Water   Hot Oil

Graph showing the steam injection system to soften up thick oil deposits

Possible Reserves – A reserve that has a chance of being developed under favorable circumstances and is given a probability of 10% to deliver the oil to the customer at an acceptable price.

So how long will those oil reserves last?

Currently published reports say there are one trillion barrels of proven reserves in the world. "Probable reserves" are reported to consist of about 5 trillion barrels. The probable oil reserves consist of hard to get to and very heavy oils that require a larger investment in equipment and technology to extract and transport.

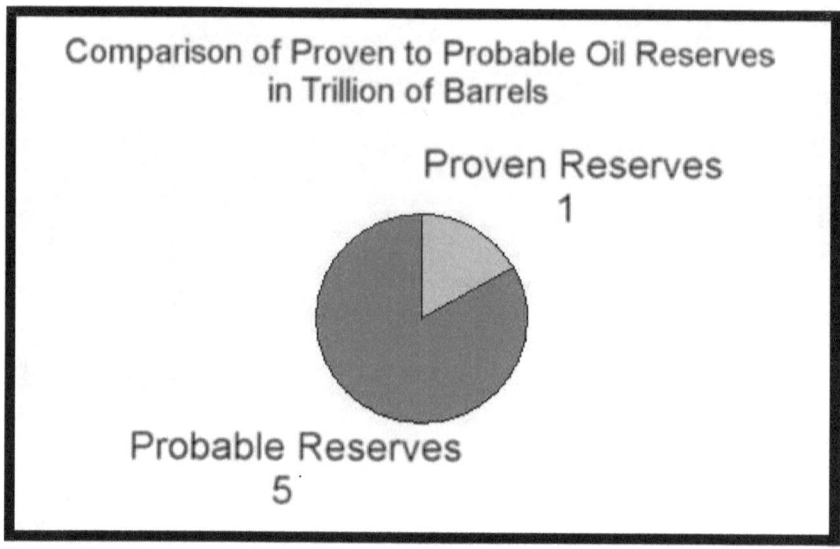

The current pumping rate of proven oil reserves is 11 billion-barrels a year. If this rate of pumping continues and the harder to get "Probable Reserves" are brought on line as well the current oil reserves will last 65 years.

Unfortunately, politicians and economists and not geologists are in control of forecasting crude oil reserves 90% of the time (such as the data above) due to its strategic importance. So in essence nobody really knows for sure how much oil is in existence. What is known is that we are discovering less and less new reserves which are essential to meet the rising demand and perhaps the price of oil is our best indication of the amount of crude oil that is available to meet that demand.

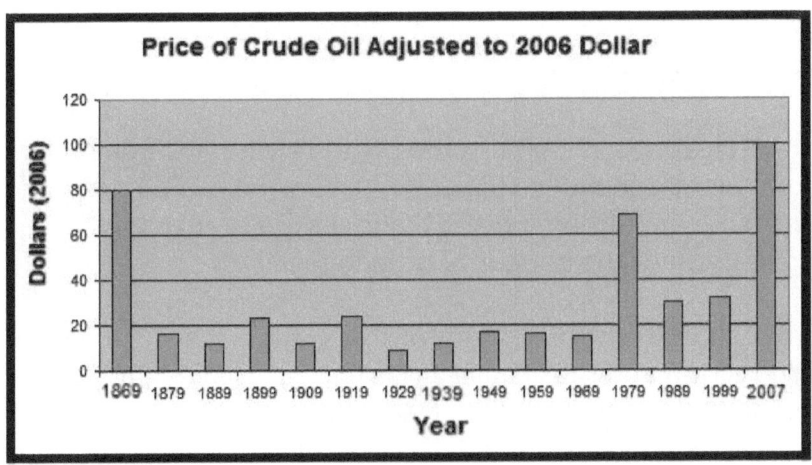

Graph showing crude oil price trends since 1869 adjusted to 2006 dollars

So how does the price of crude oil reflect reserves? That question will be explained in the next few chapters.

_____The Economics of Crude Oil

Economics is country centered meaning each country has its own economic goals and means to achieve those goals and in a global economy there are producer and consumer countries and therefore producer and consumer economics.

For example, a producing country like Saudi Arabia is highly dependent on a consumer country like the US. The standard of living in Saudi Arabia has steadily risen over the years requiring larger and larger exports of crude oil in order to generate more cash. Unfortunately, it isn't a mater selling more crude oil that increases its income of cash because more crude oil on the open market only reduces its price since its availability exceeds demand.

Fortunately for Saudi Arabia, the US' demand for crude oil has steadily increased and so Saudi Arabia has been able to grow more cash by pumping more oil to meet the ever rising demand without the danger of reducing prices.

Another course of action Saudi Arabia can take is to cut crude oil production and thus cause a rise in its price. Again, Saudi Arabia would not fair much better than increasing production because they will be selling less crude oil besides the US can switch to another source of energy if the price of Saudi Arabia crude oil is too high. In addition, an increase in the price of crude may cause inflation and possibly a recession in the US which in turn slows economic growth which is not good for Saudi Arabia because in a US recession crude oil consumption drops thus Saudi Arabia would lose revenue.

That isn't to say that there hasn't been major ups and downs for the price of crude oil over the years but those events were triggered by disruptions to the supply of crude oil due to political events.

In general, the price of crude oil has been steadily increasing faster than inflation and although that may indicate that the supply is decreasing or demand is increasing it tells us little about true crude oil reserves.

But economists are not concerned too much about true reserves of crude oil simply because if the price gets too high people will switch to a cheaper source of energy which will be discussed in later chapter on alternative sources of energy.

___Coal

Coal is found in the ground. It is valued for its energy content, and since the 1880s has been widely used to generate electricity. Steel and cement industries use coal as a fuel for extraction of iron from iron ore and for cement production.

A lot of developments over the recent years have involved coal mining, from the early days of men tunneling, digging and manually extracting the coal on carts to large open cut and long wall mines. Mining at this scale requires the use of draglines, trucks, conveyor, jacks and shearers.

Coal mining can have a huge environmental impact and needs to be managed. Many mines are required by government to rehabilitate the area that was mined.

_____Exploration

Coal must be found in a sufficient concentration and quality if it is to be economically recovered. Most of the world's mines are centered on the ancient 'shield' rocks of the Precambrian eras of 4.6 billion to 2.6 billion years ago, and 2.6 billion to 570 million years ago. This is because the mountain-building activity, which helped concentrate many minerals, was intense in this early period of the planet's life.

The stages of exploration:

Geophysical Surveys - Airborne evaluation of magnetic or density anomalies, which are good indications of areas prospective for mineral deposits.

Mapping - Consolidation of the surface expressions into a single plane for better understanding of the likely deposit configuration.

Sampling - Collection of stream sediments, surface boulders or earth (the latter usually from trenches dug across a prospective area). The material is then analyzed to test for anomalous concentrations of metals to establish drilling targets.

Grab Samples: Often the coal grab sample is not representative of the entire coal seam because it is found on the surface of the ground and is one of the least reliable coal sampling methods.

Channel Samples: When the coal sample is collected from an outcrop, the exposed area should be cleaned to avoid the weathered exposed coal surface. Normally a small box cut is made at the coal outcrop exposing the entire thickness of the coal seam. For a relatively thin seam only one coal section is recommended. However, if the seam is really thick, two or more coal sections may be necessary to sample entire seam.

Core sampling. This method involves the drilling for a core sample that gives examples of the type and quality of coal in the seam.

Cutting/Chip samples: This is a much less accurate sampling scheme than the core sampling. Cuttings are generated by rotary type drilling where no core is recovered in the form of chips. This king of sampling can only give us a very general analysis of the coal. It is very difficult to collect samples and most of the time we have lots of impurities mixed in it. Also, the exact depth of coal can't be accurately recorded unless generated from a geophysical log after drilling is completed.

Bulk Samples: Bulk samples are collected mainly for larger scale tests, to check swelling properties of various coal seams, to rank coal by High Pressure coals and low pressure coals and so on.

Drilling - Recovery, for analysis, of either rock chips (at various depths) or cores (collection of the latter is by using diamond-encrusted circular drill bits and core barrels).

Modeling - Evaluation of grades and known structures (often using computer models) to determine the likely deposit configuration.

Infill Drilling – Is the drilling of extra holes at the site to increase confidence in the model.

Feasibility Studies - Various scenarios are tested at different coal prices to determine if the deposit can be extracted profitably. The last such study is called a 'bankable' feasibility study as it is used to secure funding.

There are various classifications for ore deposits, depending upon the certainty that the configuration is understood (this is usually a function of the number of drill holes):

Surface Mining

When coal seams (a coal find is called a seam) are near the surface, it may be economical to extract the coal using open cut (or strip) mining methods. Open cut coal mining recovers a greater proportion of the coal deposit than underground methods, as more of the coal seams in the strata may be exploited. Large Open Cut mines can cover an area of many square kilometers and use very large pieces of equipment. Explosives are first used in order to break through the surface of the mining area. The top soil is then removed. Once the coal seam is exposed, it is drilled, fractured and thoroughly mined in strips. The coal is then loaded on to large trucks or conveyors for transport to either the coal preparation plant or directly to where it will be used.

Most open cut mines in the United States extract bituminous coal. In Australia and South Africa open cut mining is used for both thermal and metallurgical coals. In New South Wales open casting for steam coal and anthracite is practiced. Surface mining accounts for around 80 percent of production in Australia, while in the US it is used for about 67 percent of production. Globally, about 40 percent of coal production involves surface mining.

Strip mining

Strip mining exposes the coal by removing the top soil in long cuts or strips. The soil from the first strip is deposited in an area outside the planned mining area. Spoil from subsequent cuts is deposited as fill in the previous cut after coal has been removed. Usually, the process is to drill the strip of top soil next to the previously mined strip. The life of some area mines may be more than 50 years.

Mountaintop removal mining

 Mountaintop coal mining is a surface mining practice involving removal of mountaintops to expose coal seams, and disposing of associated mining soil in adjacent "valley fills." Valley fills occur in steep terrain where there are limited disposal alternatives.

 In areas with rolling or steep terrain with a coal seam occurring near the top of a ridge or hill, the entire top is removed in a series of parallel cuts. Top soil is deposited in nearby valleys and hollows. This method usually leaves ridge and hill tops as flattened plateaus. The process is highly controversial for the drastic changes in topography, the practice of creating head-of-hollow-fills, or filling in valleys with mining debris, and for covering streams and disrupting ecosystems.

 Top soil is placed at the head of a narrow, steep-sided valley or hollow. In preparation for filling this area, vegetation and soil are removed and a rock drain constructed down the middle of the area to be filled, where a natural drainage course previously existed. When the fill is completed, this under-drain will form a continuous water runoff system from the upper end of the valley to the lower end of the fill. Typical head-of-hollow fills are graded and terraced to create permanently stable slopes.

Room and pillar mining

 Room and pillar mining consists of coal deposits that are mined by cutting a network of rooms into the coal seam. Pillars of coal are left behind in order to keep up the roof. The pillars can make up to forty percent of the total coal in the seam.

Underground mining

Most coal seams are too deep underground and require underground mining, a method that currently accounts for about 60 percent of world coal production. In deep mining, the room and pillar or board and pillar method progresses along the seam, while pillars and timber are left standing to support the mine roof. Once room and pillar mines have been developed to a stopping point that is limited by geology, ventilation, or economics, a supplementary version of room and pillar mining, termed second mining or retreat mining, is commonly started. Miners remove the coal in the pillars, thereby recovering as much coal from the coal seam as possible. A work area involved in pillar extraction is called a pillar section.

Modern pillar sections use remote-controlled equipment, including large hydraulic mobile roof-supports, which can prevent cave-ins until the miners and their equipment have left a work area. The mobile roof supports are similar to a large dining-room table, but with hydraulic jacks for legs. After the large pillars of coal have been mined away, the mobile roof support's legs shorten and it is withdrawn to a safe area. The mine roof typically collapses once the mobile roof supports leave an area.

Modern mining

Technological advancements have made coal mining today more productive than it has ever been. To keep up with technology and to extract coal as efficiently as possible modern mining personnel must be highly skilled and well trained in the use of complex, state-of-the-art instruments and equipment. Many jobs require four-year university degrees. Computer knowledge has also become greatly valued within the industry as most of the machines and safety monitors are computerized.

The use of sophisticated sensing equipment to monitor air quality is common and has replaced the use of small animals such as canaries, often referred to as "miner's canaries".

In the United States, the increase in technology has significantly decreased the mining workforce.

Environmental impacts

Coal mining can result in a number of adverse effects on the environment. Surface mining of coal completely eliminates existing vegetation, destroys the soil profile, displaces or destroys wildlife and habitat, degrades air quality, alters current land uses, and to some extent permanently changes the general topography of the area mined. This often results in a scarred landscape with no scenic value. Rehabilitation or reclamation mitigates some of these concerns and is required by US Federal Law.

Mine tailing dumps produce acid mine drainage which can seep into waterways and aquifers, with consequences on ecological and human health. If underground mine tunnels collapse, this can cause subsidence of land surfaces. During actual mining operations, methane, a known greenhouse gas, may be released into the air. And by the movement, storage, and redistribution of soil, the community of microorganisms and nutrient cycling processes can be disrupted.

_____Production

Over 8,050 Mt/year of coal is currently produced in over 50 countries. Coal production has grown fastest in Asia, while Europe has declined. The top coal mining nations (figures in brackets are 2009 estimate of total coal production in millions of tons) are:

Production of Coal by Country in 2010
(million tons)

Country	2010	Reserve Life (years)
China	3240.0	35
USA	984.6	241
India	569.9	106
EU	535.7	105
Australia	423.9	180
Russia	316.9	495
Indonesia	305.9	18
South Africa	253.8	119
Germany	182.3	223
Poland	133.2	43
Kazakhstan	110.8	303
Total World	7,273.3	118

Most coal production is used in the country of origin, with around 16 percent of coal production being exported.

Coal reserves are available in almost every country worldwide, with economically recoverable reserves in around 70 countries. At current production levels, proven coal reserves are estimated to last 147 years. However, production levels are by no means level, and are in fact increasing and some estimates are that peak coal could arrive in many countries such as China and America by around 2030.

```
Proved recoverable coal reserves
    at end-2008 (million tons)

Country                    TOTAL

United States              237,295
Russia                     157,010
China                      114,500
Australia                   76,500
India                       60,600
Germany                     40,699
Ukraine                     33,873
Kazakhstan                  33,600
South Africa                30,156
Serbia                      13,770
Colombia                     6,746
Canada                       6,528
Poland                       5,709
Indonesia                    5,529
Brazil                       4,559
Greece                       3,020
```

___Natural Gas

Naturally occurring natural gas is a hydrocarbon gas mixture consisting primarily of methane (and up to 20 percent ethane and a small amount of impurities) Natural gas is an important energy source in many applications including heating buildings, generating electricity, providing heat and power to industry and vehicles and is also used in the manufacture of products such as fertilizers.

 It is found in deep underground natural rock formations or associated with crude oil reservoirs, in coal beds. Most natural gas was created over time by two mechanisms:

1- Biogenic gas is created by organisms in marshes, bogs, landfills, and shallow sediments.

2- Deeper in the earth, at greater temperature and pressure, gas is created from buried organic material.

_____Natural Gas Exploration

Geological Surveys

Exploration for natural gas begins with geologists examining the surface of the earth, and determining areas where it is geologically likely that gas deposits could exist. By surveying and mapping the surface and sub-surface characteristics, the geologist can figure out which areas are most likely to contain a natural gas reservoir. From the outcroppings of rocks on the surface or in valleys and gorges, to the geologic information attained from the rock cuttings and samples obtained from the digging of irrigation ditches, water wells, and other oil and gas wells are just a few of the tools available to geologists.

If the geologist has determined an area where it is geologically possible for a natural formation to exist, further tests can be performed to gain more information.

Seismic Exploration

The Earth's crust is composed of different layers. Each layer has its own properties. For example, energy traveling underground interacts differently with each of these layers. These seismic waves, emitted from a source, will travel through the earth, but also be reflected back toward the source by the different underground layers.

Seismic waves are created artificially. The reflected waves are then picked up by sensitive pieces of equipment called 'geophones' that are embedded in the ground which are then transmitted to a seismic recording truck, for further interpretation by geophysicists and petroleum reservoir engineers. The source of seismic waves can be an underground explosion or thumping by a thumping truck as described in the section on crude oil exploration.

Offshore Seismology

When exploring for natural gas that may exist thousands of feet below the seabed floor, a ship is used to pick up the seismic data while hydrophones are towed behind in various configurations depending on the needs of the geophysicist. The seismic ship uses a large air gun, which releases bursts of compressed air under the water, creating seismic waves that travel through the Earth's crust and generate the seismic reflections.

Magnetometers

The magnetic properties of underground formations can be measured to generate geological and geophysical data by using magnetometers, which are devices that can measure the small differences in the Earth's magnetic field.

Gravimeters

Different underground formations and rock types all have a slightly different effect on the gravitational field that surrounds the Earth. By measuring these minute differences with very sensitive equipment, geophysicists are able to analyze underground formation.

Exploratory Wells

This consists of digging into the Earth's crust to allow geologists to study the composition of the underground rock layers in detail. In addition to looking for natural gas and petroleum, geologists also examine the drill cuttings and fluids to gain a better understanding of the geologic features of the area.

2-D Seismic Interpretation

Geophysicists using the data collected from seismic exploration activities to develop a cross-sectional picture of the underground rock formations.

Computer Assisted Exploration

Computers are used to compile and assemble geologic data into a coherent 'map' of the underground.

3-D Seismic Imaging

Three-D imaging utilizes seismic field data to generate a three dimensional 'picture' of underground formations and geologic features using advanced computers. This allows the geophysicist and geologist to see a clear picture of the composition of the Earth's crust in a particular area.

4-D Seismic Imaging

This type of imaging is an extension of 3-D imaging technology where changes in structures and properties of underground formations are observed over time.

_____Storage and transport

It is not easy to store natural gas or transport by vehicle. Natural gas pipelines are impractical across oceans. Many existing pipelines in America are close to reaching their capacity. In Europe, the gas pipeline network is already dense in the West. New pipelines are planned or under construction in Eastern Europe and between gas fields in Russia, Near East and Northern Africa and Western Europe.

Gas is turned into liquid at a liquefaction plant, and is returned to gas form at regasification plant at its destination. liquefied natural gas (LNG) is transported across oceans in tanker ships, while tank trucks can carry liquefied or compressed natural gas (CNG) over shorter distances.

LNG is the preferred form for long-distance, high-volume transportation of natural gas, whereas pipeline is preferred for transport for distances up to 4,000 km over land and approximately half that distance offshore.

CNG is transported at high pressure, typically above 200 bars. Compressors and decompression equipment are less capital intensive than LNG equipment and may be economical in smaller unit sizes than liquefaction/regasification plants. Natural gas trucks and carriers may transport natural gas directly to end-users, or to distribution points such as pipelines.

Natural gas is often stored underground inside depleted gas reservoirs from previous gas wells, salt domes, or in tanks as liquefied natural gas. The gas is injected in a time of low demand and extracted when demand picks up. Storage nearby end users helps to meet volatile demands, but such storage may not always be practicable.

Floating Liquefied Natural Gas (FLNG) is an innovative technology designed to enable the development of offshore gas resources that would otherwise remain untapped because due to environmental or economic factors it is nonviable to develop them via a land-based LNG operation.

Many gas and oil companies are considering the economic and environmental benefits of Floating Liquefied Natural Gas (FLNG). However, for the time being, the only FLNG facility now in development is being built by Shell,[36] due for completion in around 2017.

_____Natural Gas Use

Before natural gas can be used as a fuel, it must undergo cleansing to remove impurities in order to meet market specifications for quality.

The by-products of cleansing include ethane, propane, butanes, pentanes, and higher molecular weight hydrocarbons, hydrogen sulphide, carbon dioxide, water vapor, and sometimes helium and nitrogen.

With 15 countries accounting for 84% of the worldwide extraction, access to natural gas has become an important issue in international politics, and countries vie for control of pipelines.

Power generation

Natural gas is used for electricity generation through the use of gas and steam turbines. Most grid power plants and some off-grid engine-generators use natural gas. Particularly high efficiencies can be achieved through combining gas turbines with a steam turbine in combined cycle mode. Natural gas burns more cleanly than other hydrocarbon fuels, such as oil and coal, and produces less carbon dioxide per unit of energy released. For an equivalent amount of heat, burning natural gas produces about 30% less carbon dioxide than burning petroleum and about 45% less than burning coal. Combined cycle power generation using natural gas is thus the cleanest source of power available using hydrocarbon fuels, and this technology is widely used wherever gas can be obtained at a reasonable cost.

Domestic use

In much of the developed world it is supplied to homes via pipes where it is used for many purposes including natural gas-powered ranges and ovens, natural gas-heated clothes dryers, heating/cooling, and central heating. Home or other building heating may include boilers, furnaces, and water heaters.

Compressed natural gas (CNG) is used in rural homes without connections to piped-in public utility services, or with portable grills.

CNG is a cleaner alternative to other automobile fuels such as gasoline and diesel. As of 2008 there were 9.6 million natural gas vehicles worldwide, led by Pakistan (2.0 million), Argentina (1.7 million), Brazil (1.6 million), Iran (1.0 million), and India (650,000). The energy efficiency is generally equal to that of gasoline engines, but lower compared with modern diesel engines. Gasoline vehicles converted to run on natural gas suffer because of the low compression ratio of their engines, resulting in a cropping of delivered power while running on natural gas (10%–15%). CNG-specific engines, however, use a higher compression ratio due to this fuel's higher-octane number of 120–130.

Hassan Rasheed

An Introduction to Energy

Uses

___Engines

_____The Human Muscle

You might consider a muscle something other than a machine but in reality a muscle runs on hydrocarbons, it produces carbon dioxide like a car's, and it involves the movement of parts much like that of an "engine."

A muscle is a bundle of many elongated cells called fibers. Muscle fibers are quite small. They are from about 1 to 40 microns long and 10 to 100 microns in diameter. A typical cell in your body is about 10 microns in diameter, and one strand of hair is about 100 microns.

A muscle fiber contains many smaller fibers called myofibrils, which are cylinders of proteins which do the actual contraction of muscles. Myofibrils contain two types of filaments. There are thick (made of a protein called myosin) and thin filaments (made of another protein called actin). Each thick filament is surrounded by six thin filaments.

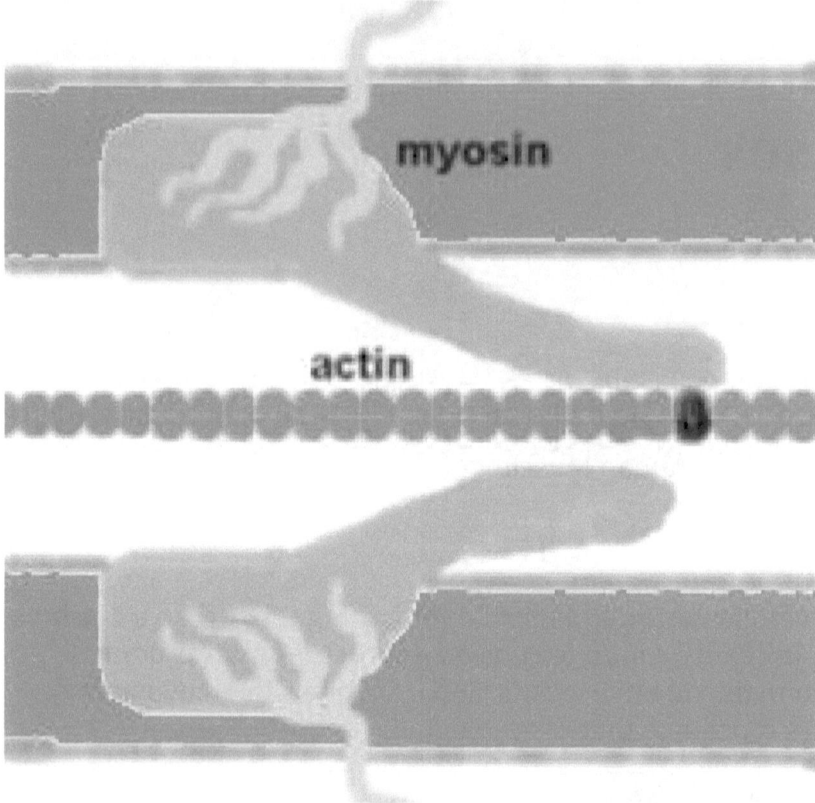

During contraction, the myosin thick filaments grab on to the actin thin filaments. The thick filaments pull the thin filaments past them in a fashion much like what you would do when pulling a rope by alternating your grip on the rope while pulling with your right hand then your left.

The energy used to fuel this mechanism is called adenosine tri-phosphate (ATP). From its name you can see that it is a molecule with three phosphate atoms. The process of releasing energy from this molecule requires the separation of one of the phosphate atoms and transferring the energy that held it in place to the muscle cell to do work. The result is a molecule called adenosine di-phosphate (ADP) and a free atom of phosphate.

The trigger that makes a muscle contract is a nerve signal. This consists of an impulse made up of Calcium atoms that are released by the nerve cell membrane after it is stimulated.

The source of ATP is an ADP molecule and a phosphate atom. These come together using energy from the bodies energy storage device called fatty acids. Fatty acids consist mostly of long chains of carbon and hydrogen like fatty acids shown in a previous section.

Each muscle contains enough muscle fibers to do the job required by the body. The need to do more or harder work triggers the body to produce more muscle to meet those needs. When the need no longer exists the body absorbs or metabolizes muscle fibers so essentially the body regulates the amount of muscle to meet the needs of the body to do work. This principle is expressed across living creatures to conserve energy.

So what kinds of apparatus or machine does an economy run on?

_____The Heat Engine

Economies also require devices to make it run and the modern economy requires for the most part a device called the heat engine.

All combustion engines can be classified as "Heat engines" because they depend on a heat deferential to convert energy into mechanical work. Not all heat engines are equally efficient at converting fuel to mechanical energy. Some require more energy and thus emit more carbon dioxide to produce the same output of work. The following chart compares the most popular modes of transportation to the amount of carbon dioxide they produce.

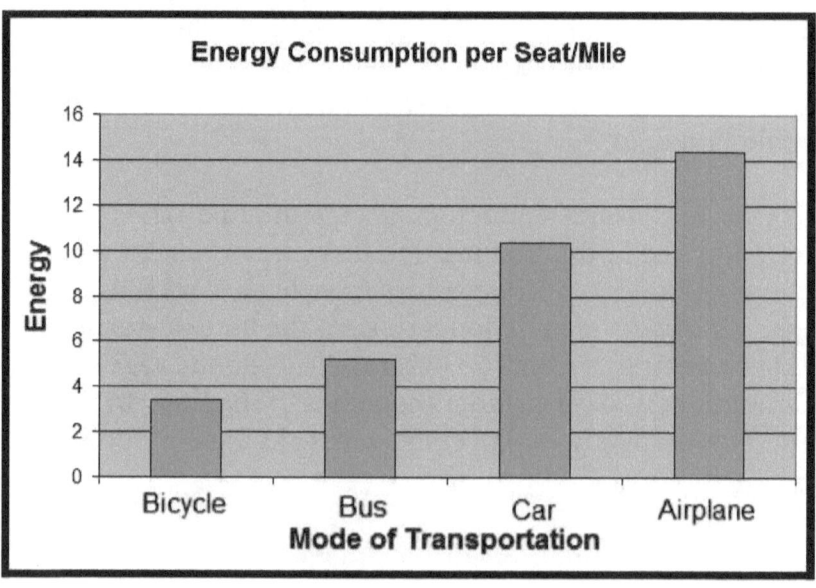

Graph comparing different modes of transportation and their relative consumption of energy

The above chart shows 'seat/Mile" data. A seat-mile is a way of standardize quantities between vehicles that seat one or many individuals at the same time. What it describes is the energy consumed for each seat no matter how many seats are in any particular vehicle. The calculation takes the total energy consumed for a vehicle traveling one mile and divides that figure by the number of occupied seats in that vehicle.

Please note I have included the bicycle in the above chart for comparison reasons. It is true that all of the above methods of transportation consume energy but the bicycle depends on respiration for its energy and thus can't be classified as a heat engine.

Heat engines include the following machines:

> Car engines

> Diesel engines

> Coal fired plants

> Airplane engines

> Nuclear power generators

You might ask, "Why does a person driving a car or traveling by airplane consume more energy than a person riding a bicycle?" The answer will become apparent in the next few chapters.

Motion

On Earth we are constantly dealing with "friction." Friction is a force that hampers motion and so when you roll a ball on an even surface it eventually stops. At this point we will stop talking about friction and it is enough to have it in the back of your mind for the time being.

Any object that is in motion has energy associated with it. For example, a bullet traveling through the air has more energy than one resting on the ground. If we have an energy meter

in its path we can measure its energy of motion when it slams
into the meter. As it turns out the energy of the moving bullet
is related to its mass (a term related to its weight) and its
velocity (or speed).

To use some numbers loosely, the energy content of the
moving bullet is half its mass times its velocity squared (or
roughly half its weight times its speed times its speed). So if a
bullet weighing two ounces is traveling at the speed of 1000
feet per second it would have the motion energy of 1,000,000
units. If it weighed four ounces then its motion energy
content would be 2,000,000 units. If it weighed two ounces
but traveling at 2000 feet per second then its motion energy
content would be 4,000,000 units.

Looking at it in another way, if there is a bullet weighing two
ounces sitting in the chamber of a gun and we want to launch
it into the air at a speed of 2000 feet per second then there
must be enough gun powder in the bullet shell to generate or
deliver 4,000,000 units of energy.

Friction (resistance)

On earth, the bullet described above would fall from the sky
after about a mile or so of travel. This is because the shape of
the bullet causes it to plow through atoms in the air which
slow it down to the point that it falls from the sky to the
ground.

To describe the process of plowing through air atoms and
coming to a standstill we can say that in order for a bullet
traveling with 4,000,000 energy units through the air that
comes to a stop must have encountered 4,000,000 resistance
or friction units. In other words, the energy that went into

giving the bullet its speed is equal to the energy of resistance that brought it to a stop.

Internal Combustion Engine

The most common type of heat engine is the internal combustion engine meaning that its source of energy is combustion of a fuel in an internal chamber to generate heat. The heat of combustion is converted into mechanical energy that drives the auto. By a spark from the spark plug the benzene and oxygen molecules in the fuel mixture from the carburetor are converted to molecules of carbon dioxide and water in the piston plus mechanical energy (movement of the piston).

Benzene ring

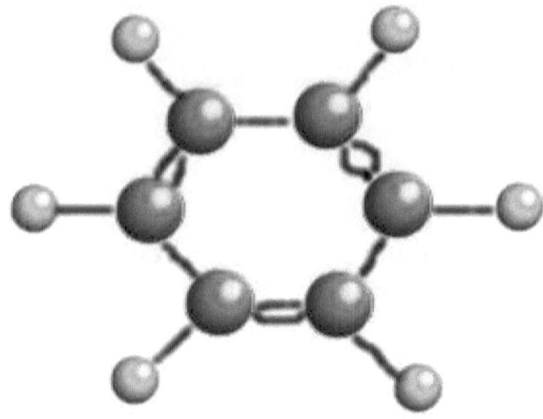

 Hydrogen

Carbon

Graph showing the atoms and their relationships of a benzene molecule

The way the internal combustion engine works is best explained by watching the action of one piston. A piston is a plunger like device that travels up and down the inside of a cylinder made of steel closed at one end. With the plunger starting at the closed end, the exhaust valve closes and the plunger starts to move out of the steel tube. As it does, another valve at the closed end is opened (the fuel-air valve) and fuel mixed with air is sucked in.

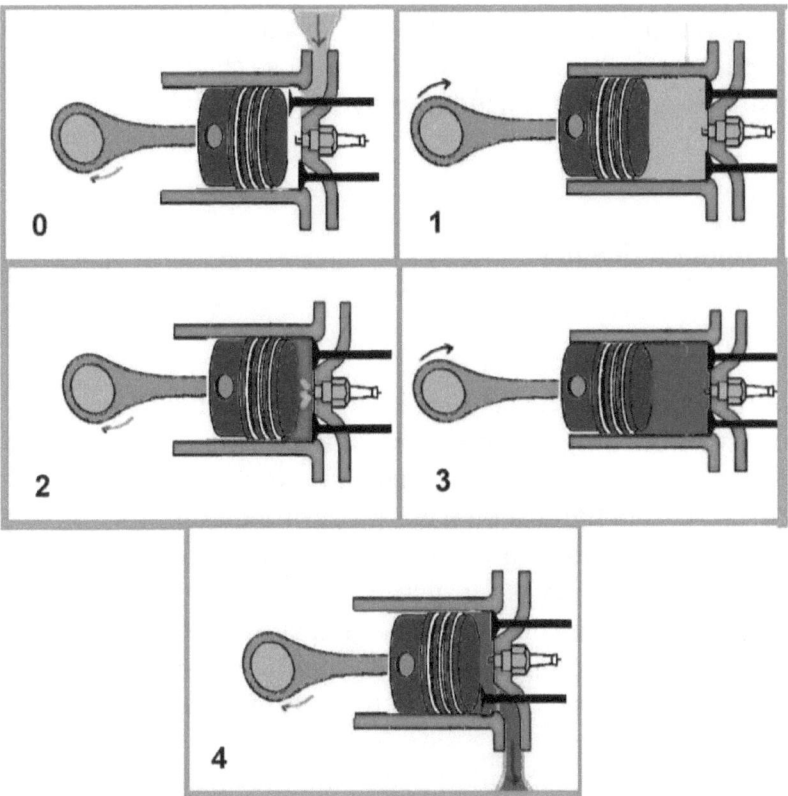

Graph showing the different strokes of an internal combustion engine

The fuel-air valve is closed when the plunger reaches the open end of the steel tube and the plunger then reversed course and starts to move towards the closed end of the steel tube compressing the fuel-air mixture. When the plunger reaches the closed end a spark is delivered that ignites the compressed fuel-air mixture forcing the plunger to move back to the open end of the steel tube. The inertia of the engine pushes the plunger back to the closed end and the exhaust valve opens allowing the consumed gas out.

An Introduction to Energy

I have described so far four "strokes" of the plunger the third of which is called the power stroke because the fuel is burned releasing energy that is used to move the plunger out.

Heat Engine Efficiency

Present technology has achieved a maximum of 40% efficiency in burning the fuel. That means for every 100 units of energy that are consumed by the auto engine 40 units are converted to mechanical energy. Most of the 60 units that are not converted to mechanical energy are dissipated as heat and removed by the cooling system. It is these four strokes of the engine that takes place at very high speeds that causes the internal combustion engine to be so inefficient. The formula for why this is will be covered in the next few pages.

Thus the auto engine is called a heat engine and it depends on the difference between the temperature inside the piston and the temperature of the air outside the engine. The greater the difference in temperature the greater amount of energy is converted to mechanical motion. An ultimate engine would have a piston chamber temperature infinitely high and the temperature outside the engine infinitely low.

Unfortunately the materials used to maintain an engine don't allow these extremes in temperatures. For example, the oil that is used to lubricate the pistons would burn or disintegrate and cause piston damage if the temperatures got too hot.

Other factors that affect energy transformation in autos are the gas-air mixture that enters the piston and the degree that the fuel-air mixture is compressed.

How exactly does the energy get used up?

Let us start simply. First, consider what it takes to travel a mile. The average person burns about 600 calories of fuel walking that mile and their efficiency is about 16%.

A single mile on a bicycle will required the burning of about 400 calories of fuel. Although the bike weighs about 40 pounds the bicycle conserves momentum: once it starts rolling, it keeps on rolling.

The internal combustion engine is a whole different story. It takes 20,000 calories of fuel to move the whole 4000 pound vehicle that one mile.

There are three factors that influence fuel consumption. They are:

The internal combustion engine is 20% efficient.

The faster you go the more wind and road resistance there is.

The longer the distance you want to go, and the shorter the time in which you want to get there, the more fuel is consumed.

The influence of speed increases exponentially as you go faster and faster. What this means is that you will spend about three times more energy going two miles in the time it took you to go one mile. At first you will be using a lot of energy in accelerating from a standstill and once you reach your speed wind and road resistance come into play causing you to expend more energy.

This relationship between time and distance caused some auto makers to place an RPM (Revolutions per Minute) gage in their autos. This gage times the number of revolutions per minute of the crank shaft which has a direct relationship to the turning of the wheels and thus the auto's speed.

In actuality, the efficiency of an auto with an internal combustion engine is much less than 20% because the process of gaining the speeds (acceleration) that people feel they need, requires revving up the engine (increasing its RPM) and resulting in greater carbon dioxide emissions.

The following three diagrams illustrate the effects of size of auto, speed and acceleration on fuel consumption.

First, the size or volume of an auto is important to fuel consumption because of drag. Drag is caused by many factors that include the size of the engine, conditions of the road to lubrication. In the following diagram we look at the effects of the size of the auto. The large model is a Hummer and the small model is a Honda Civic. Please note that the chart describes fuel consumption in the form of carbon dioxide emissions since an increase in fuel consumption is an increase in carbon dioxide emissions and an indication of greater energy use.

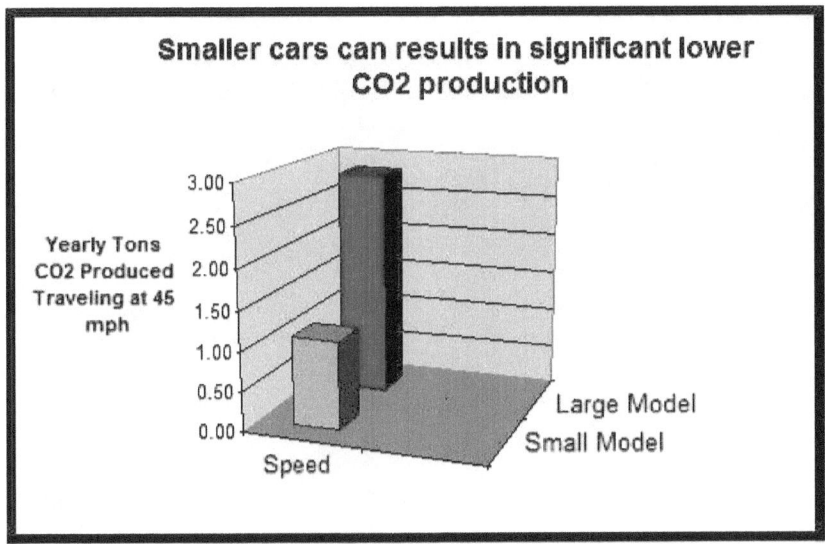

Graph showing the relative levels of emissions of Carbon dioxide between a small and large auto

In the next diagram we look at a reduction in speed of a Hummer by 10 miles per hour and the effect on carbon dioxide emissions.

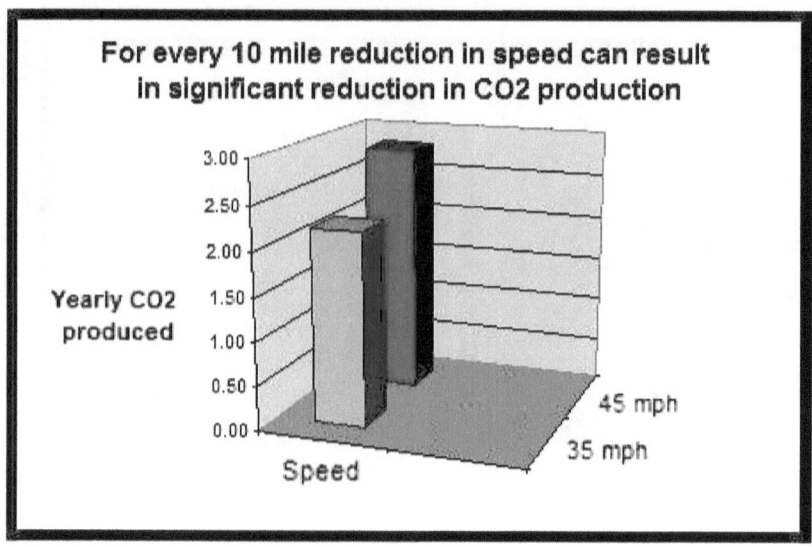

Graph showing the relative levels of emissions of carbon dioxide between using the same auto at different speeds

The following third diagram shows the effects of increasing the RPM from 1000 to 2000 by increasing the acceleration of an auto and the resultant increase in fuel consumption. Note that the effects of drag are over shadowed by the effects of the engine trying to speed up.

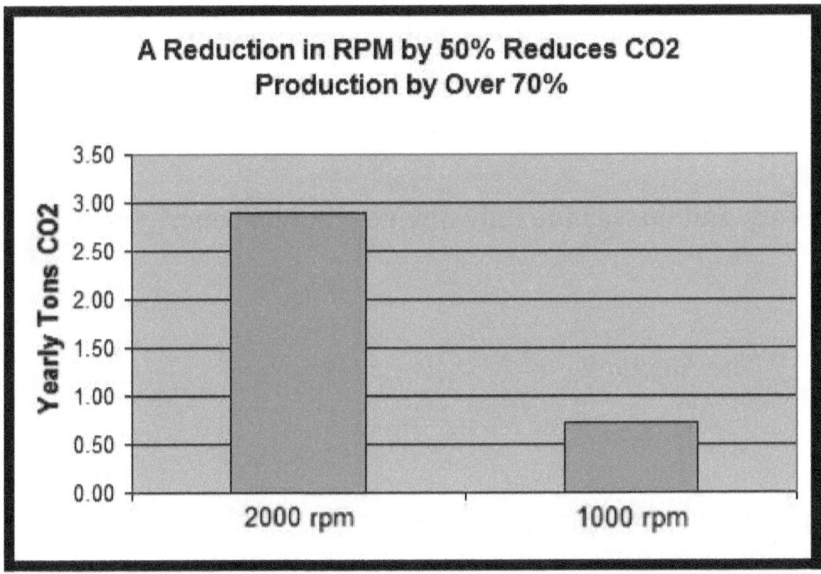

Graph showing the effects of different RPM (Rounds per Minute) of an auto engine and its effects on carbon dioxide emissions

In summary, the cost of going that one mile requires the engine to consume 80% of its energy simply because it exists. The remaining 20% are consumed overcoming drag and achieving the desired speed through acceleration.

Principle: 1

It may be useful to us to make a comparison between the cost of making an automobile and the cost of operating it. It takes about 100-500 million calories to manufacture an auto and in its lifetime it will use 500 billion calories in gasoline. So the operation of an auto or a machine is generally a source of more energy consumption by about 3,000 times of manufacturing it. So when purchasing a machine, its fuel or operational cost on the environment overshadows its initial cost.

Are cars and busses the only devices that use energy?

_____The Airplane

Airplanes use the jet engine and the Turbofan engine is the most common form of jet engine in use today. It is used in airliners like the Boeing 747 and military jets, where an afterburner is often added for supersonic flight. The First stage compressor is greatly enlarged to provide bypass airflow around the engine core.

Compared to other jet engines the turbofan is quieter due to greater mass flow and lower total exhaust speed. It is more efficient for a useful range of subsonic airspeeds resulting in cooler exhaust temperatures.

As far as its consumption of oil, it is the greatest of all the machines we have talked about and thus very inefficient. The cost of traveling at speeds of 500 miles per hours greatly overshadows any gains in reduced resistance and drag through air stream design and altitude.

Is the heat engine the only consumer of energy in our economy?

_____Electric Generators

If you ask, does electricity generate carbon dioxide emissions, the answer is yes.

Electric generators consist of a magnet fastened to an axel that revolves inside of an insulated wire coil. As the magnet revolves it generates an electric current in the coil. The two ends of the coil are then used as contact points for other electrical devices such as the plug in for your refrigerator.

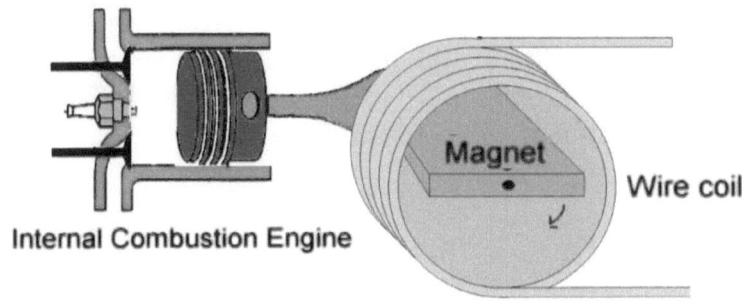

Internal Combustion Engine   Magnet   Wire coil

Graphic caricature of the different parts of an electric generator

The problem with electric generators is that they need a source of power to generate power. Diesel engine-generator sets operated at their best efficiency point can produce 3.5

kilowatt hours of electrical energy for each liter of diesel fuel consumed (and 2 pounds of carbon dioxide) (efficiency = 30%), with lower efficiency at partial load.

It is estimated that electricity generates 2.7 billion tons of carbon dioxide emissions a year through the use of electricity generated by oil, natural gas and cola fired electric generators.

Principle 2:

You can never achieve 100% efficiency when converting energy from one form to another. Further, the efficiency of a series of conversions of energy is only as efficient as the product of the individual efficiencies. For example, an electric generator consists of a series of two power conversion steps. The first is the conversion of diesel fuel to mechanical power (maximum of 40% efficiency). Second, the mechanical power is converted to electrical power (maximum of 80%). The result is a maximum of 40% times 80% which gives us 30% efficiency.

Coupled with the fact that less efficiency means more fuel consumption, a less efficient process produces greater quantities of greenhouse gasses. So a process that generates electricity at 30% efficiency produces 70% more greenhouse gasses.

_____Coal Fired Plants

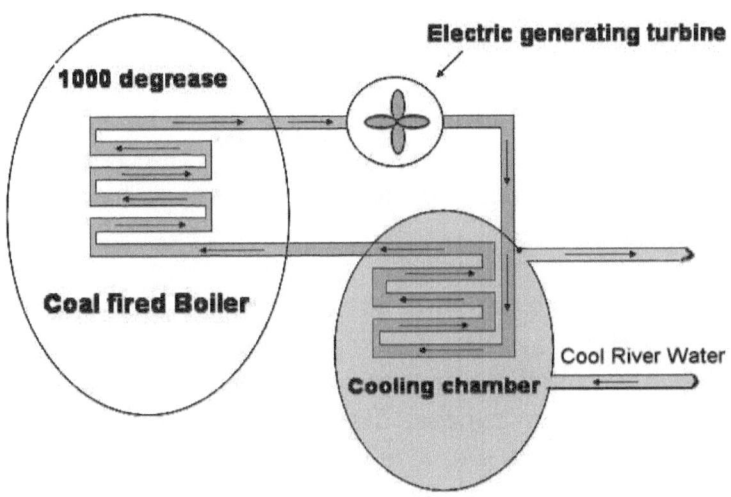

Graph of the different parts of a coal fired plant

The coal fired boiler plant is also a heat engine and is based on the design of the old locomotive steam engines where water is converted to steam in a boiler fired by coal and the steam drives a piston to drive the wheels but in the case of the current electric generating plants the steam drives a turbine that in turn drives an electrical generator.

The boiler raises the temperature of the water to 1000 degrees and the cooling chamber is usually cooled with river water.

An Introduction to Energy

Most of the coal in the US is used to generate electricity and the industry produces 2.2 billion tons of carbon dioxide emissions a year.

_____Electric Motors

Electric motors consume about 60% of the electricity generated in the US at the current time and that percentage will increase as alternative sources of energy are brought into the mainstream.

Its principle design depends on magnets to work. Magnets are devices that have two poles; one positive and the other negative. If you bring a positive pole from one magnet close to a negative pole of another they will attract one another. If you bring two negative or two positive poles together they repel one another.

These attracting and repelling characteristics are at the heart of how the electric motor achieves from 50% to 90% efficiency. They are designed so that at any one moment similar poles are close to one another thus creating the force to generate mechanical energy.

The basic design includes an electro-magnetic magnet tied in its middle to a rotating axel and at least two stationary magnets place at opposite ends and perpendicular to the axel. The magnets are placed in such a way so that when the axel rotates its magnet passes within close proximity to them.

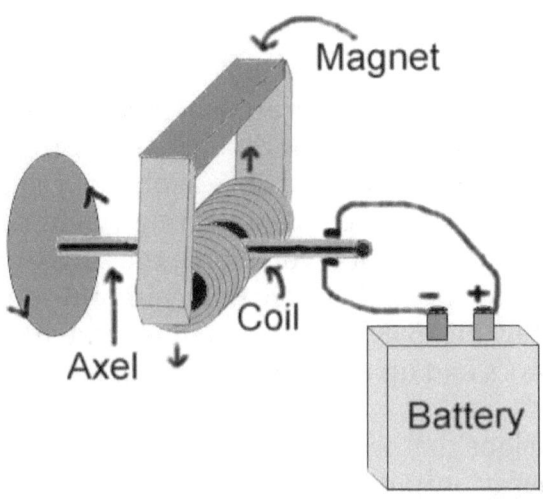

Graph showing the different parts and their relationships of an electric motor

An electro-magnetic magnet is made up of a coil of wire wound around an iron rod or core. When electricity passes through the coil the iron core becomes a magnet. Depending on the direction of the flow of electricity, an electro-magnetic magnet will have two pole states (positive and negative) and the design of the axel is such that as it rotates the direction of electricity flow switches direction and thus switching the poles in the magnet.

The switching of poles is made at the time the magnet on the axel is in close proximity to the stationary magnets and with an opposite polarity so that a repelling force is present and repels the rotating magnet away around the axel. Batteries are connected to the axel in such a way as to switch the direction of flow as the axel rotates. Thus the axel will continue to rotate until the batteries run out of power or are disconnected.

The discrepancy between the 50% and 90% efficiency rating is due to the way that motors are matched to the work that they do. If you match a motor that generates 100 units of energy of work and you match it to a job that requires 90 units of energy then the efficiency rating would be approximately 95%. But if you match the same motor to a job that requires 50 units of energy then your efficiency would approximately be 75%. Overall, the miss match of jobs to electric motors end up wasting 24-40% of electric energy.

Many of the jobs that we ask of electric motors require variable energy output which most can't do. There are electric motors that have variable power output but they are generally more expensive to purchase and thus they suffer the same symptoms of alternative energy sources discussed so far in that most consumers go for the least expensive initial installation costs even though in the long run on going operational costs are far more expensive.

_____The Battery

An important part of the previously discussed zero Carbon Dioxide emissions technologies is the battery. It is used to store electrical energy and deliver it when it is needed. So when the sun is up, the wind is blowing or when the ocean is restless, we are able to generate power and charge batteries for the times that they are not.

Electricity is the flow of electrons through a conductive path. This path is called a circuit. Many dissimilar metallic elements will generate a current through a closed circuit. An example is the earphone radio that operates by connecting two wire leads to different metallic objects like lead and copper plumbing pipes.

Batteries have three parts, a negative anode, a positive cathode, and an electrolyte between them. The cathode and anode (the positive and negative nodes at either end of a traditional battery) are hooked up to an electrical circuit generating an electrical current to do work.

The electrolyte causes a chemical reaction in the battery, and electrons build up in large numbers, causing an imbalance between the anode and the cathode. The electrons want to rearrange themselves to get rid of this instability or difference in charge. By connecting the anode and cathode through a closed circuit, the electrons migrate to the cathode, creating a current that can do work.

These electrochemical processes change the chemical composition in the anode and cathode rendering them less capable of performance, and the battery's efficiency gradually declines.

When you recharge a rechargeable battery, you reverse the directional flow of electrons using another power source, such as solar panels. The electrochemical processes happen in reverse, and the anode and cathode are restored to their original state and can again provide full power. The electrolyte, anode and cathode parts of a rechargeable battery are made of different materials than a non-rechargeable battery.

Rechargeable batteries have an efficiency of around 75% to 85%. This efficiency is measured by dividing the output power by the power used to recharge it, and generally don't reflect the cost of the battery itself.

An Introduction to Energy

Batteries reach peak performance in power delivery when they are fully charged. This performance declines as it discharges. Heavy duty or deep cycle batteries maintain high performance even when they are half or a quarter of the way charged. Deep cycle batteries are used when the application can only charge them up part of the time, such as with solar power cells.

Batteries need to have a minimum charge, also known as a "seed charge" in order to be recharged effectively. Batteries also have a relative short life span of about five years due to their chemical composition, and proximity of the elements that constitute it. These changes accelerate when the battery is less than fully charged.

Due to the chemical nature of batteries, they should to be disposed of properly or recycled when possible when they stop being useful, to avoid releasing the many toxic and corrosive compounds they contain. This adds to their overall cost and the cost of the applications they comprise.

Hassan Rasheed

An Introduction to Energy

Impact

___Global Warming

Introduction

Let's examine what a lot of scientists are talking about today namely "Global Warming." There are a few facts that have caught their attention. One is the shrinking of the polar ice caps. This is accompanied by a warming trend in global temperatures as can be seen in the following chart.

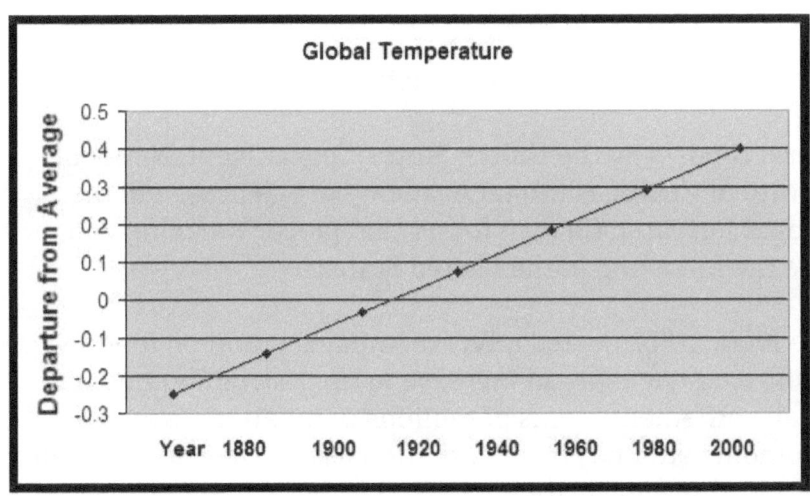

Graph showing the trend in global temperatures

It shows the departures of the global temperatures from its average. As you can see there is an upward trend. It also shows that temperatures have risen about 0.7 of a degree since the 1880's. And so we have a cause and effect of the melting of the polar ice caps.

What scientists also realize is that if this trend in warming continues for the next fifty to one hundred years huge chunks of the northern and southern ice caps will disappear and the consequences of which will be a significant rise in ocean water levels that endanger our coastal communities and environments.

What is not clear is what is causing the temperature to rise to such a level as to cause all these changes. Today, some scientists are focusing on the release of greenhouse gases a major component of which is carbon dioxide. Other scientists indicate that the Sun is getting hotter and therefore causing an increase in global temperatures.

Looking into the scientific crystal ball

Several reports by the United States Government, National Institute of Health, National Academy of Sciences, and the National Research Council found that global warming will cause the following in the United States:

More Floods: "Projected adverse impacts based on models include . . . a widespread increase in the risk of flooding for human settlements (tens of millions of inhabitants in settlements studied) from both increased heavy precipitation events and sea level rise."

Increased spread of infectious diseases: "an increase in the number of people exposed to vector borne diseases (e.g. cholera) and an increase in heat stress mortality."

Degraded water quality: "Projected climate change will tend to degrade water quality through higher water temperatures and increased pollutant load from runoff and overflows of waste facilities."

More frequent and more intense heat waves, droughts, and tropical cyclones: "The vulnerability of human societies and natural systems to climate extremes is demonstrated by the damage, hardship, and death caused by events such as droughts, floods, heat waves, avalanches, and storms."

Great Lakes Water Level Decline

"Climate change is likely to reduce water levels in the Great Lakes areas and river levels in the central US during the summer, thereby affecting navigation and general water supplies."

Snow pack Decline

"It is very likely that as the climate warms, less precipitation will fall as snow, the existing snow pack will melt sooner and faster, the runoff will be shifted from late spring and summer to late winter and early spring."

Increased Floods

"Climate change is likely to increase flood frequency and amplitude in some regions, with major impacts on infrastructure and emergency management."

___The Controversies

There are many theories as to why the temperatures around the globe are rising. Some theories say that it is part of the natural cycle of geologic events in the solar system while others say that it is manmade. Here we will examine both.

Global warming as a natural occurrence

A major theory supporting the natural occurrence of global warming is that the sun goes through cycles of glowing hotter, then cooler every seventy thousand years or so. As a result there are such events on Earth as period of extreme heat and other periods of extreme coolness as exhibited by the occurrence of successive Ice Ages in North America and Europe and Asia. This theory suggests that we are currently in a period of global warming after the last global cooling off period that resulted in the last Ice Age that ended ten thousand years ago.

A case in point is related to the Sphinx of Egypt. Unlike many of the artifacts found in Egypt, the period and builders of the Sphinx are unknown. But one study of the stones of that monument show water marks suggesting floods or strong torrential rains which are unheard of in Egypt today with its arid and hot climate. When archeologists are pressed for a time period in which the Sphinx was built they suggest ten thousand years ago.

So here we have two data points to consider. They are the end of the last Ice Age with periods of torrential rains in Egypt about ten thousand years ago and today a hot and arid climate there. If you connect the two data points on a two-dimensional graph of time vs. temperature you can see that we are in a warming trend that predates the industrial revolution and the advent of manmade greenhouse gasses.

Global warming as a manmade phenomenon

Every chemical has a range in the electromagnetic spectrum at which it absorbs energy from a photon coming from the sun. The sun emits photons with energy wave lengths of around 400 nanometers. These photons travel through the atmosphere until they reach whatever is covering the earth's surface.

What isn't absorbed by the surface of the earth is reflected back at wavelengths less than 400 nanometers. Carbon dioxide, methane and nitrous oxide are three of the chemicals in the atmosphere that absorb light at wave lengths lower than 400 nanometers.

When a greenhouse gas absorbs this low wavelength photon it stars to vibrate and may or may not release its energy to the surrounding environment in even lower wavelengths of heat.

But carbon dioxide, methane and nitrous oxide are not the only gasses that exhibit this phenomenon. Almost any gas can absorb photon energy and release it to the surrounding environment as heat.

One of the gasses that exhibit this phenomenon is water vapor and in reality, it can cause over 95% of the greenhouse effect. So why aren't environmental scientists clamoring over water vapor as a Green House Gas.

The reality is that without the ability of gases in the atmosphere to absorb and release energy at lower (heat) wavelengths is essential to the livability of the planet Earth. Otherwise, it would exhibit a climate like that of the surface of the moon and would be uninhabitable.

So, gasses like water vapor, carbon dioxide, Methane and nitrous oxide are essential to make our Earth livable with warm temperatures that are more or less within a livable range night or day season to season and what the environmental scientists are concerned about is any excess in these gasses that seem to be correlated to a global rise in temperatures and the dangers posed by it.

It is these excesses in gasses that are over and beyond what environmental scientists consider essential to a stable and livable environment and are termed Green House Gasses.

In the following chapters we will learn about the origins of these Green House gasses and how they are related.

Conclusion

Although the increased presence of greenhouse gasses (GHG) in the atmosphere did not give us a cause-and-effect connections yet, they do raise the suspicions of most scientists working on global warming. The simple fact that our environments will change dramatically in the next hundred years or so leaves us wondering what we can do and so we come back to the experiments in the laboratory that tell us that if we control the concentration of carbon dioxide, methane and nitrous oxide in the air we can control, to a degree, the temperature whether greenhouse gasses are the culprit or not.

In other words, if it's getting hot we have the option to take of our jackets.

___Green House Gasses

In the laboratory

If you prepare a large caped glass flask with ordinary air and a green piece of filter paper inside of it and expose it to the rays of the sun the temperature inside of it will rise due to the different gasses in the air.

If you add pure carbon Dioxide to the above flask the temperature inside of it will rise more quickly and become warmer. If you add pure methane you will find that the warming effect is 21 times that of Carbon Dioxide and similarly if you add pure Nitrous Oxide the warming effect is 330 times that of Carbon Dioxide.

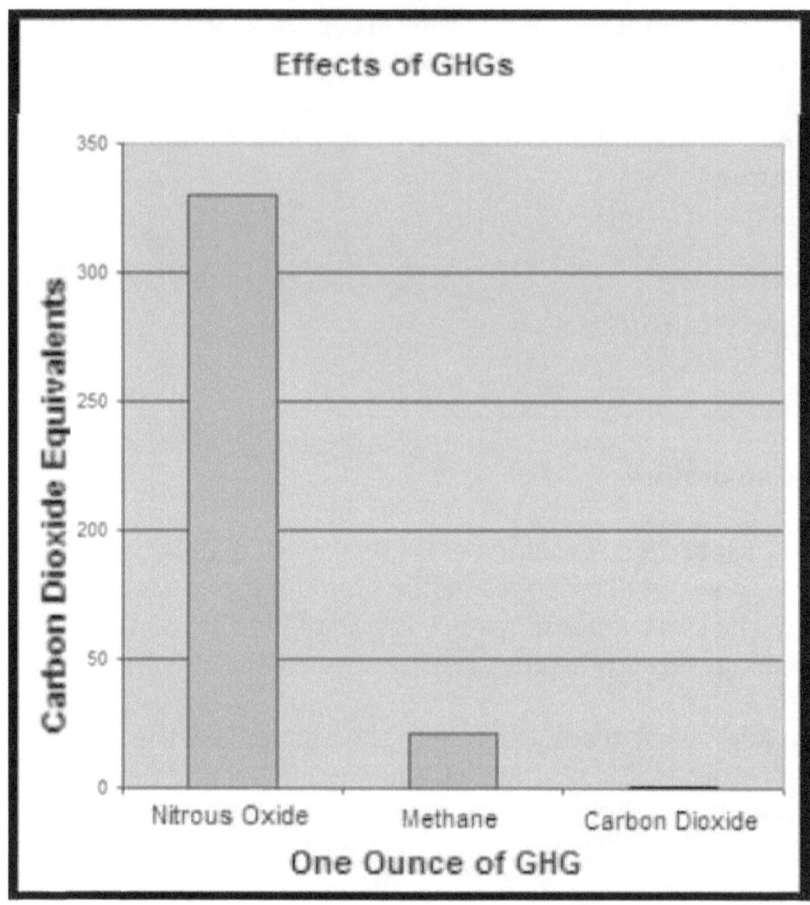

Graph showing the effects of different greenhouse gasses on temperature in controlled conditions in the laboratory

If you start with two capped glass flasks with ordinary air and green pieces of filter paper in them, add a little water to each of the green filer papers and to one add greenhouse gasses the water will evaporate with greater speed in the flask with the greenhouse gasses leaving the filter paper dryer.

Similarly, if you add ice to both flasks, the ice in the flask with the greenhouse gasses will melt much faster than the ice in the air only flask.

Given the above experiments, what are scientists saying will happen in the next 25 to 100 years?

The Predictions

Many scientists think that increased evaporation could result in more extreme weather as global warming progresses. Global average water vapor concentration and precipitation are projected to increase by the second half of the 21st century.

With an increase in water vapor in the air there will be more clouds and in turn it is likely that precipitation will have increased over northern mid- to high latitudes and Antarctica in winter. At low latitudes there are both regional increases and decreases in rain fall over land areas. Larger year to year variations in precipitation are very likely over most areas where an increase in mean precipitation is projected

Other scientists predict that lower latitude glaciers will melt away and the polar ice caps will partially melt starting at lower latitudes especially in the Western Artic and Island. Glaciers hold water, release it through the warmer months and the process of glacier melting will cause erosion, flooding and landslides at first then when they are gone draught conditions with negative impact on the environment including agriculture.

The melting of the polar icecaps will lead to an increase in the volume of water in oceans causing coastal areas to experience a rise in water levels that may reach 5 to 12 feet, depending on the expert quoted, by the middle of the 21st century making the recent devastation in New Orleans and hurricane Katrina a minor incident in comparison.

So, what is the evidence that any of this is happening?

Today we are at the very early stages of global warming and its effects but there are signs that we are on the road to major changes in weather and ocean patterns.

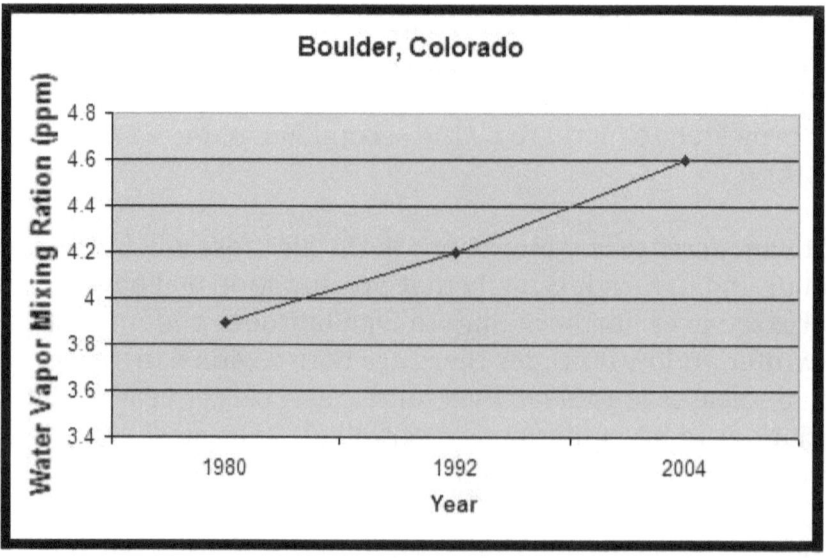

Graph showing actual concentration of water vapor in the air above the city of Boulder, Colorado

Indeed, the water vapor content of the air is rising as can be seen in the above graph showing an average increase in water vapor of .04 parts per million per year over Boulder Colorado.

And indeed, lower latitude glaciers are melting away to the point that only 50% of them are left in the world since the year 1900.

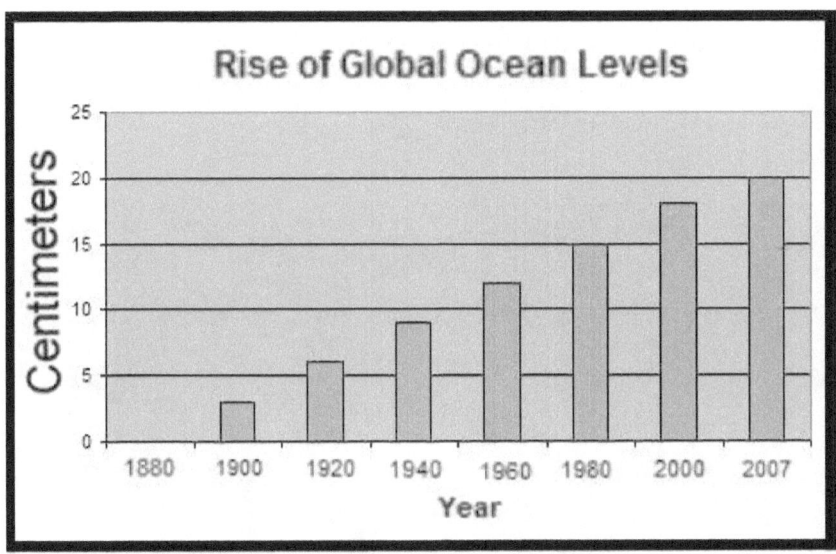

Graph showing actual sea level changes over the last one hundred years

In addition, the ocean and sea levels are now rising at the rate of an inch every ten years and is now accelerating.

In the United States forest fires are burning larger sections of land and are reaching higher up the mountains where there was once snow to slow them down. These fires are also burning hotter and trees that depend on fires are dying from the extreme heat the fires generate such as the Ponderosa Pines.

___Carbon Dioxide

It is an odorless and tasteless gas made up of one carbon atom and two oxygen atoms.

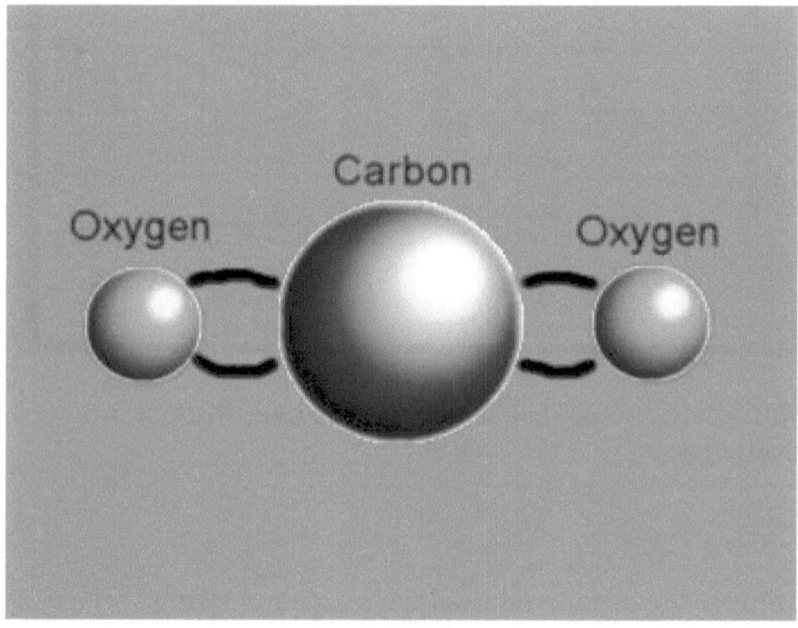

Graph of a carbon dioxide molecule

It exists as one of a group of gasses in the atmosphere at a concentration of .038% by volume. Its sister gases include Nitrogen (78.084%), Oxygen (20.949%), Argon (0.934%), water vapor (>1%), and a group of other gases (0.002%).

It weighs 1600 kilo grams per meter cubed. It has a melting point of minus 57 degrees centigrade and a boiling point of minus 78 degrees centigrade. Its solubility in water is 1.45 kilo grams per meter cubed.

Carbon dioxide also exists in other places such as the oceans and underground. There is a balance of carbon dioxide between the atmosphere and the oceans meaning that the atmosphere and the oceans exchange carbon dioxide all the time depending on the chemical composition of each as well the their temperatures.

Scientists like to talk about carbon dioxide cycles which are how these exchanges take place. One major cycle is the photosynthesis and respiration cycles of the living matter on Earth. In plants photosynthesis takes carbon dioxide and water and produces sugars and starches (carbohydrates) that all living matter depend on for energy. In respiration the reverse happens where oxygen and carbohydrates are combined releasing carbon dioxide and water once again.

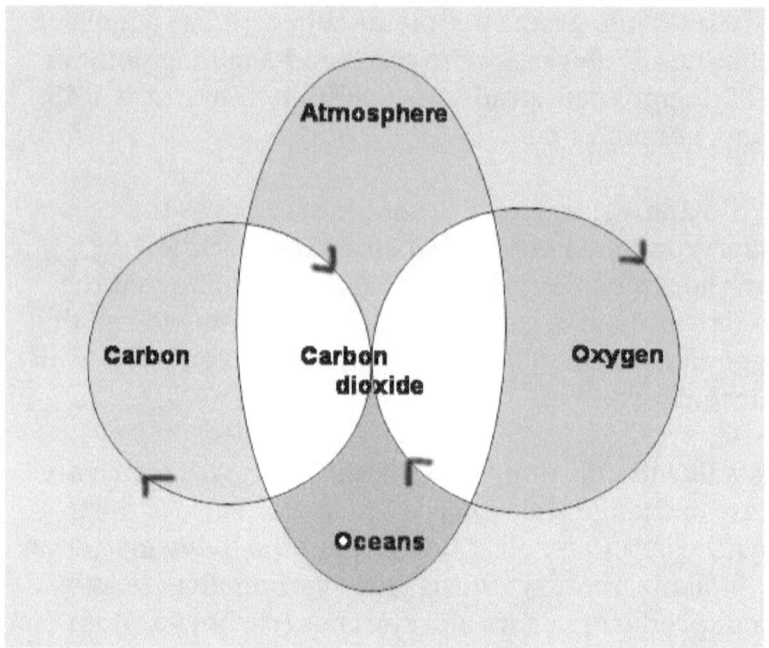

Graph showing an example of the carbon dioxide cycles

The general idea of a cycle is one that explains how a chemical move through the environment staring in one part of it then going to anther then back again where it started.

Before the advent of drilling for oil or mining for coal to fuel the steam engine and subsequent internal combustion engines the percent of carbon dioxide in the atmosphere had lowered and the reason is a constriction at one end of the carbon dioxide cycle because of geologic activities millions of years ago that had buried large quantities of carbohydrates (fossil fuels) and prevented their content of carbon dioxide from returning to the atmosphere.

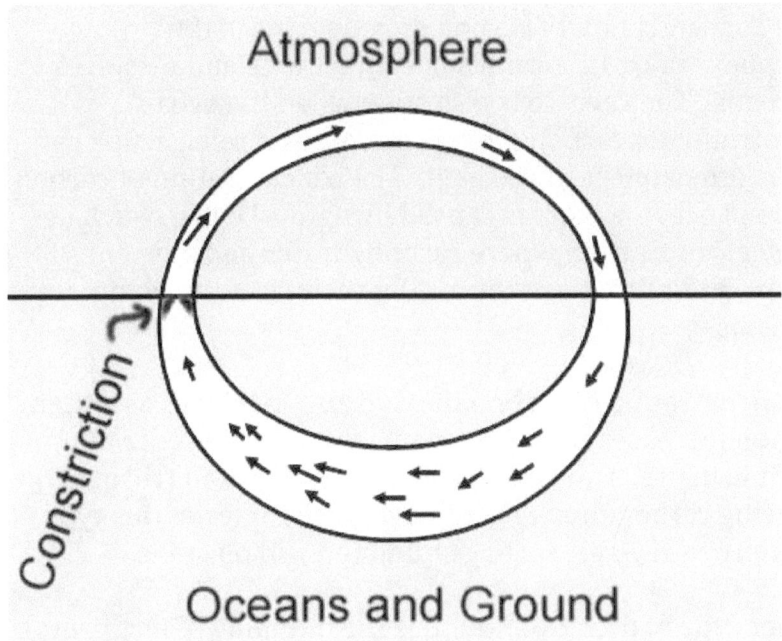

Graph showing an example of a constricted natural cycle

After the advent of the industrial revolution the blockage of the carbon dioxide cycle was mitigated through coal mining, oil drilling, natural gas exploration and the concentration of carbon dioxide is returning to what it was millions of years ago.

The balance of carbon dioxide between the atmosphere and the oceans

Carbon dioxide is in constant flux between the oceans and the atmosphere. Ocean waters consist mainly of water (H_2O), dissolved sodium chloride salt and a host of other chemicals including organic matter, oxygen and carbon dioxide.

An Introduction to Energy

As the concentration of carbon dioxide rises in the atmosphere more of it comes in contact with and dissolves in the oceans. The reverse is also true; when the ocean's concentration of carbon dioxide increases it release this gas into the atmosphere once again. The concentration of carbon dioxide in ocean waters is caused by natural events such as biological respiration where carbohydrates and oxygen combine and release energy for the various ocean going organisms.

The concentration of carbon dioxide also increases as water pollution increases because the major component of pollution is organic matter and chemicals that are digested (respired) producing carbon dioxide and water. Examples of this type of pollution are sewage, garbage dumping and oil spills.

Another type of ocean pollution is the dumping of fertilizers that cause algae blooms followed by die offs and decomposition which also produced carbon dioxide.

Other conditions that affect the oceans ability to hold carbon dioxide besides concentration is its temperature. Cold waters hold more carbon dioxide (or gasses in general) than warm ones.

And thus as the atmosphere warms up and in turn so does the oceans, the oceans give up more of its carbon dioxide and compounds it atmospheric greenhouse affects.

The carbon dioxide cycles described above are part of the greater carbon cycle (shown in the following diagram.) For example, when we talk about respiration carbon dioxide starts out in the form of carbon-hydrogen chains that are combined with oxygen to produce carbon dioxide and water. So technically carbon dioxide is the output of this respiration process and its origin is a carbohydrate. So essentially, we have carbon atoms that in one state are combined with oxygen and in another state are combined with hydrogen and so it is more accurate to describe such processes as part of the carbon cycles.

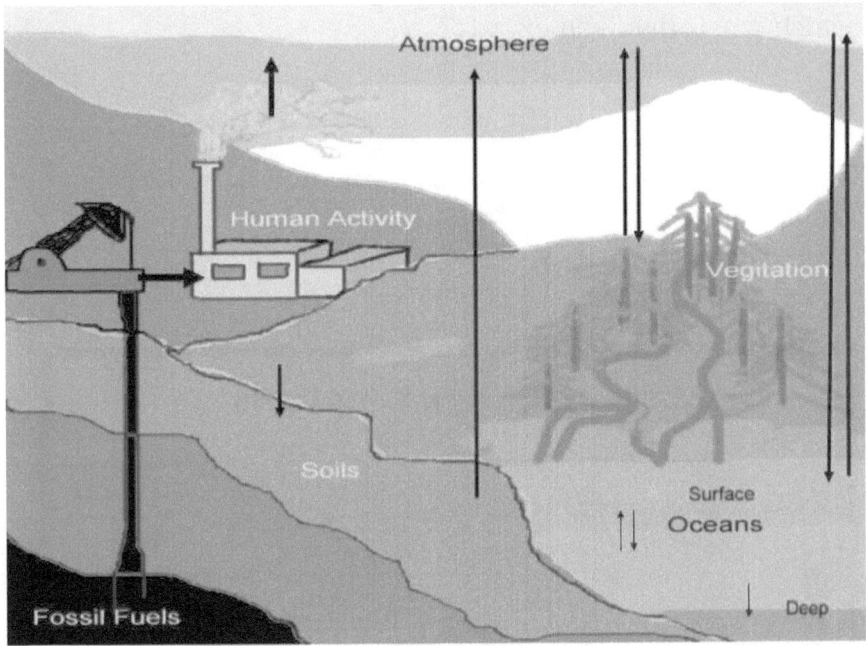

Graph showing one representation of carbon cycle

So how is the United States involved in the emission of carbon dioxide?

The U.S.A. and Carbon Dioxide

Today the US is pumping 6.1 billion tons of carbon dioxide into the atmosphere from those ancient carbohydrate deposits. But these deposits consist of oil, gas and coal so which one is involved exactly?

The answer is from all three as can be seen from the following graph.

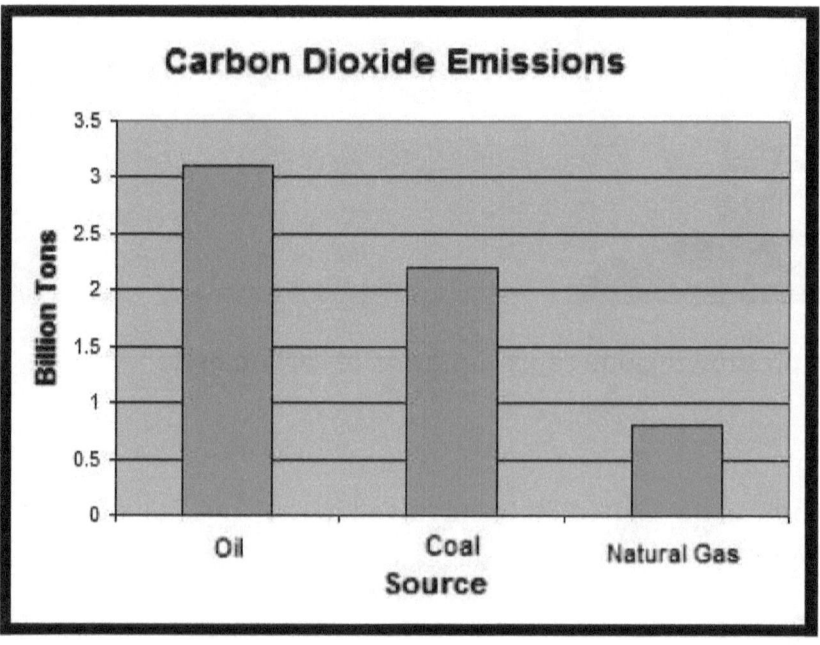

Graph of the major contributing chemical sources of carbon dioxide

The major raw materials contributors to carbon dioxide production are oil, coal, and natural gas in that order. As it turns out most carbon dioxide is released through internal combustion engines and coal fired plants as can be seen from the following chart.

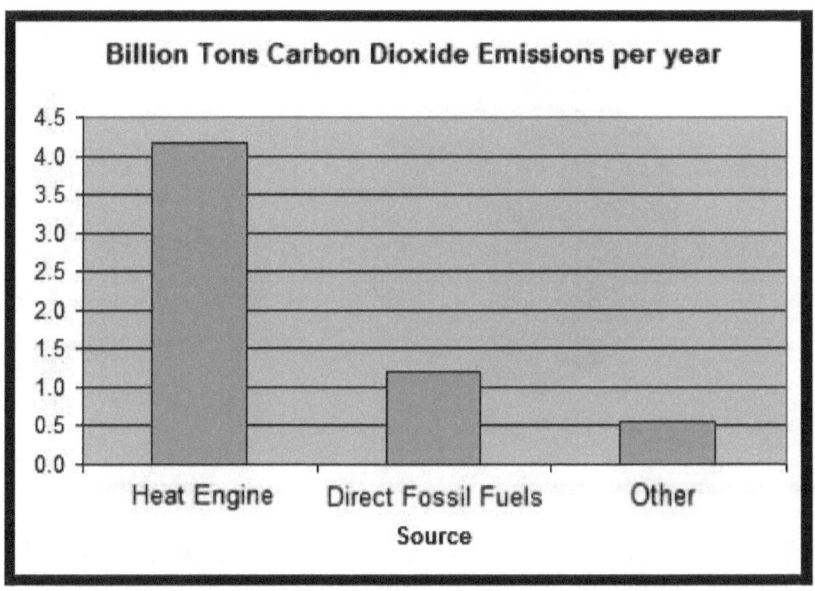

Graph of the sources of combustion that produce carbon dioxide

Other sources of carbon dioxide include direct use of fossil fuels in such activities as burning, and manufacturing. As of this study there is about half a billion tons produced in the USA that we have not been able to pinpoint its application source.

To look at the above data in a different light and to show the impact that you and I have on carbon dioxide production the following chart demonstrates how much carbon dioxide is produced by residents and supporting industry in the USA.

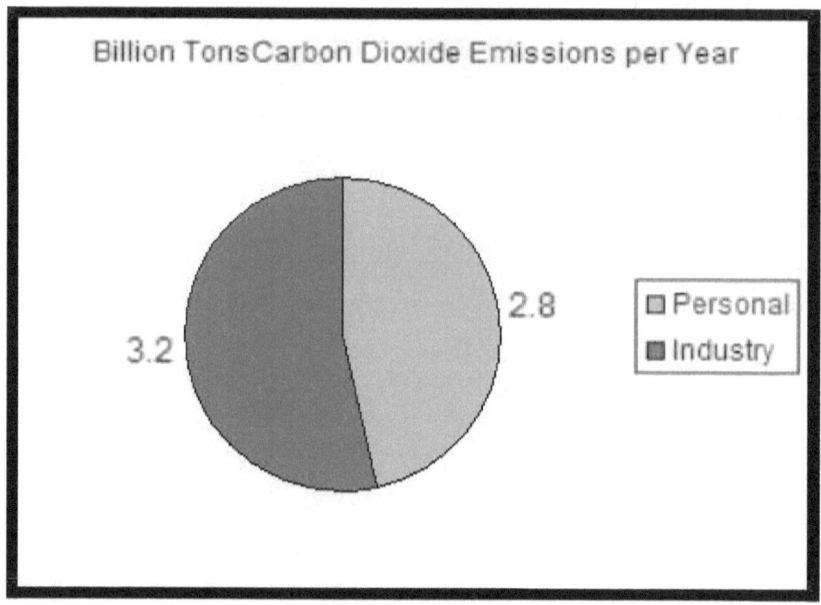

Graph of the quantitative relationship between personal and industrial sources of carbon dioxide emissions

As far as the contributions from residents of the USA are concerned, all carbon dioxide produced is from the internal combustion engine, electricity (which originates from internal combustion and fuel fired engines,) space and or water heating.

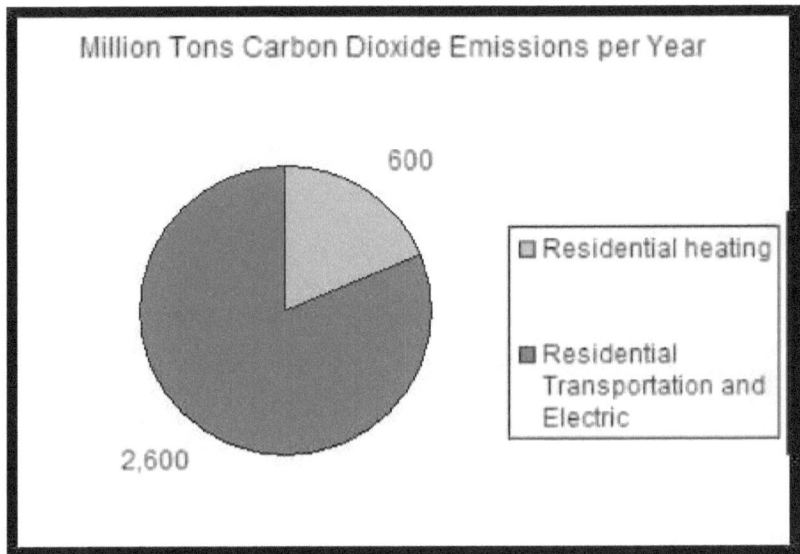

Graph of the relationship between the carbon dioxide emissions from heating and other residential activities

To take our impact at home on carbon dioxide production still further, let us examine the production of household appliances of this gas. The USA census has defined a family as consisting of 3.15 people and if we analyze their effect in producing carbon dioxide as far as appliances are concerned we come up with the following table (your family may vary depending on the number of individuals and frequency of use.)

Appliance	Pounds carbon dioxide a year
Computer - Desktop	1,696
Air Conditioning	1,413
Refrigeration	1,210
Space Heating	892
Water Heater	804
Lighting	777
TV - 25"	565
Lawn Mower	418
Electric Blanket	201
Microwave: Large	188
Dishwasher	141
Microwave: Compact	113
Oven	105
Hair Dryer	79
Toaster	58
Printer	38
Washing Machine	31
Coffee Maker	25
Frying Pan	19
Hot Plate	19
Vacuum Cleaner - Full size	9
Waffle Iron	8
Sink Garbage Disposal	7
Coffee Pot	6

Table showing different household appliances, their usage and emissions of carbon dioxide for the average family of 3.15 individuals

___Methane

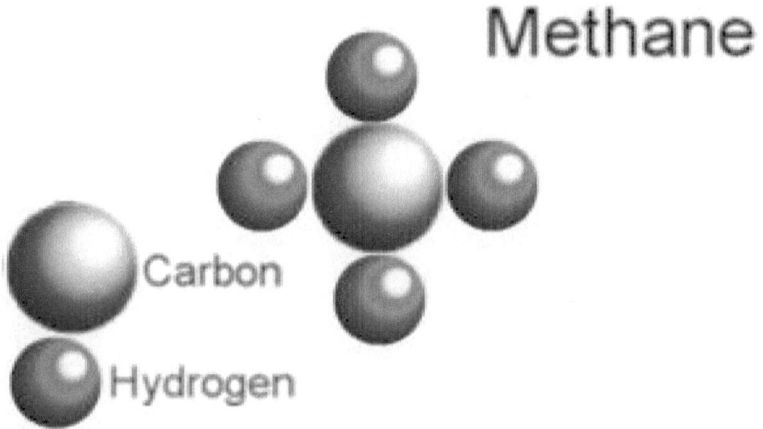

Graph showing the different atoms and their relations in a methane molecule

Methane is a gaseous compound made up of one carbon and four hydrogen atoms. It is generated from wet organic sources that are also lacking oxygen in past geologic times or in the present.

The presence of organic sources and water causes decomposition by microbes and in the presence of oxygen the decomposition is allowed to complete the oxidation of hydrocarbons to carbon dioxide and water. In the absence of oxygen the decomposition proceeds but the end-result is the simplest of hydrocarbons including methane.

An Introduction to Energy

The term aerobic decomposition refers to the process of decomposition that is done in the presence of oxygen while the term anaerobic decomposition refers to the process done in the absence of oxygen.

Temperature also has an effect on decomposition. Warmer climates produce methane at a faster rate than colder ones.

Natural sources of methane include wetlands, the digestive process of termites, oceans and hydrate deposits. The total global production of these sources is estimated at 190 million metric tons of carbon dioxide equivalents a year.

Human sources of methane in the US

Source	Million Tons
Landfills	170
Natural Gas	130
Livestock	120
Coal Mining	82
Manure	30
Waste Water	25
Petroleum	20
Other	30

Table showing the different sources of methane that are manmade and their relative levels of emissions in millions of tons of CO2 equivalents

Anaerobic decomposition

Landfills produce 130 million metric tons of carbon dioxide equivalents of methane a year and the majority of methane in the US. A landfill will produce methane if it contains a large amount of organic material and at the same time the degree of its wetness. The more organic material it has and the wetter it is the more and faster it will produce methane.

Livestock also produce methane in their digestive systems especially ruminant livestock such as cattle, sheep, camels, goats and buffalo that have multiple stomachs. The rumen is the forward stomach of these animals and is where pre-digestion takes place including fermentation which is a form of decomposition. The high content of their feed mixed with water and other digestive fluids causes the anaerobic fermentation that releases methane through the mouth of the animal. Methane is also released through the rest of the digestive process. Methane from these sources accounts for 116 million metric tons carbon dioxide equivalents of methane per year.

Further, the waste products of digestion from livestock, manure, are another source of methane if not handled properly. On large farms such as swine farms and dairies much of the waste is held in ponds and tanks which are rich in organic material and water but deficient in oxygen leading to anaerobic decomposition and methane production. These sources are estimated to emit 38 million metric tons of carbon dioxide equivalents of methane per year.

A similar problem occurs in human waste water transmission and treatment systems. Most sewer systems are underground preventing oxygen from the decomposition process leading to methane production. The treatment plants that treat the sewage an-aerobically to remove solids, chemical and microbial contaminants also release sizable quantities of Methane if the methane is release into the air. It is estimated that 35 million metric tons of carbon dioxide equivalents of methane are released each year.

Mitigation of methane emissions from anaerobic decomposition

Obviously, a reduction in the size and number of landfills will help in a significant reduction on methane production. This can be done by a more aggressive effort in recycling waste products. And more significantly is a reduction in waste in the first place. In addition, keeping landfills dry will also reduce the decomposition process and in turn methane emissions.

Limiting our consumption of beef, pork and poultry to FDA recommended portions will reduce significantly the amount of methane. This will have a two-pronged effect in reducing methane emissions through reducing the number of animals and animal waste.

Change human waste water treatment to aerobic processes to prevent the emissions of methane. If that isn't possible, cap all treatment plants and capture their methane emissions and use it as a fuel for heat engines or the heating of homes, municipal and business facilities.

_____Fossil fuel processing

Another source of methane is the result of excavating and processing of fossil fuels. The biggest culprit is the natural gas industry where methane is one of its products. Methane losses occur in the discovery, transmission, holding and delivery systems of natural gas accounting for 130 metric tons carbon dioxide equivalents of methane per year.

Coal mining also emits a significant amount of methane gas into the air from cavities in the coal mine itself or from soils that are disturbed as part of the coal mine digging process. This industry is estimated to emit 60 million tons of carbon dioxide equivalents of methane per year.

40 metric tons carbon dioxide equivalents of methane are produced from rice cultivation, abandoned coal mines, petrochemical production, iron and steel production, agricultural residue burning and the operation of heat engines.

Oil exploration and processing also produces about 18 million tons or carbon dioxide equivalents of methane per year.

___Nitrous Oxide

Nitrous Oxide

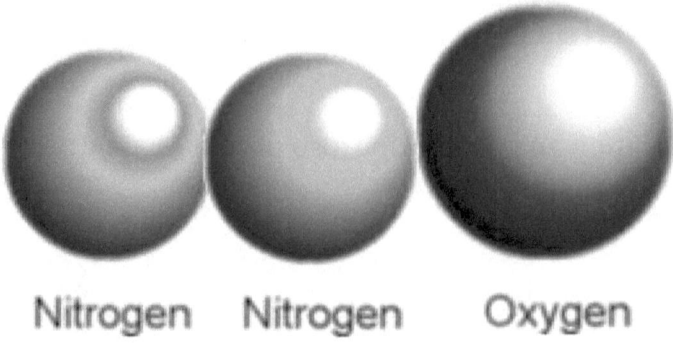

Graph showing the different atoms and their relationships in a nitrous oxide molecule

Nitrous Oxide is a gas made up of two atoms of nitrogen and one atom of oxygen. It has an elongated shape with the two nitrogen atoms bound together tightly and the oxygen atom situated to one side.

Many compounds that make up plants and animals contain nitrogen. For example, chlorophyll is what makes plants green in color and it contains four nitrogen atoms as part of its compound structure. Chlorophyll is the compound that captures the sun's photons and makes the energy captured from these photons available to other compounds and plant structures to synthesize glucose, starch and cellulose.

In animals, proteins and DNA also contain nitrogen as part of their compound structure and so do many other compounds essential for life. One function of proteins is the building of muscle fibers to assist in mobility.

When plant matter dies and falls to the ground, microbes break it down releasing the nitrogen into the environment and one of the compounds that is released is the gas nitrous oxide. Similarly, when animals defecate or when they die their nitrogen finds its way back into the environment as nitrous oxide and other nitrogen compounds.

To summarize, where there is a wet environment conducive to microbial growth that is also rich with dead plant and animal material there is production of nitrous oxide. It is therefore no surprise that 70% of the naturally produced nitrous oxide is from lush and wet tropical rain forests, tropical savannahs and the oceans. The remaining 30% is produced in temperate zones on Earth, nutrient rich estuaries and ocean sediments.

The nitrogen in nitrous oxide is naturally reincorporated into plant and animal tissues as part of the nitrogen cycle which includes nitrogen fixing plants like legumes as well as rain water that brings nitrogen compounds back to the soil for new plant and animal growth.

Scientists like to talk about the process of nitrification as being the process of incorporating nitrogen into plant and animal life and de-nitrification as the process by which nitrogen is released once more into the environment.

Natural nitrification and de-nitrification have been in balance and the amount of nitrous oxide in the environment has been relatively the same until human activity around the globe started to pick up. Records started to be kept on the concentration of nitrous oxide since 1976 and the following graph demonstrates its rapid rise in concentration in the atmosphere.

_____Human Sources of nitrous oxide

Source	Million Tons
Agriculture	275
Heat Engine	60
Nitric Acid	20
Sewage	15
Other	30

Table showing the mad made sources of nitrous oxide and their relative levels of emissions

Agriculture

At first, fingers were pointed to the auto as the reason for the increase of nitrous oxide in the environment but attention quickly turned when it was discovered that 70% of all non-natural sources of nitrous oxide comes from agricultural land management practices and the use of nitrogen fertilizers and soil conditioners to promote plant growth.

Fertilizers come in many forms. Some are manufactured and others are products like animal manures and compost. All three are high in nitrogen compound concentration and what isn't used by the crops leach into the soil where microbial action produces nitrous oxide.

Once the crop is harvested the stalks and leaves that are rich in nitrogen compounds are sometimes tilled under as a soil conditioner allowing microbes to break them down releasing nitrous oxide. Another practice is the production of animal feed from the left over crop material which in turn is converted to manure by the animals.

Nitrous oxide is also produced by crop residue burning. This practice is justified by farmers to get rid of possible plant diseases or insects that may infect subsequent crops.

Another source of nitrogen is the cultivation of legumes. Legumes are beans, vetch or soy and are known as nitrogen fixing plants meaning that they actively take nitrogen from the environment and incorporate it into its tissues including the roots. When the crop is harvested or when the plant is tilled under, the nitrogen becomes available to subsequent crops that are planted in the same area.

Nitrous oxide is also produced from the cultivation of organically rich soils. Left undisturbed, these soils produce little nitrous oxide but when tilled and irrigated for crop production their chemistry changes allowing microbes to process their nitrogen releasing it in the form of nitrous oxide.

Mitigation of nitrous oxide emissions in agriculture

Nitrogen is essential to agriculture simply because all life needs it to grow and produce. What we can do is produce more with less nitrogen.

Instead of the usual methods of broadcasting fertilizers they could be applied directly to the root of the plant much like drip irrigation is done in areas where water is in little supply. Drip irrigation delivers just the right amount of water to the root of the plant to the required depth.

Adapt drip irrigation techniques more prevalently in agriculture. One of the key components to nitrous oxide emissions is water or moister so by limiting irrigation waters we will reduce nitrous oxide emissions.

We can limit our consumption of beef, swine and poultry to the FDA's recommended portions. It takes fifty pounds of grain to produce one pound of beef or swine. By cutting our over consumption of meat we can reduce the number of acres dedicated to its production and in turn reduce the production of nitrous oxide by 25%.

With the reduction in fertilizer broadcasting, limiting the use of water and hold out appetites to smaller portions of meat we can significantly reduce nitrous oxide emissions.

_____Nitric acid production

Nitric acid is used in the manufacture of agricultural fertilizers as well as explosives for the military and industry. A byproduct of its manufacture is nitrous oxide which is released into the air through vents in the manufacturing facilities.

Mitigation of nitrous oxide emissions from nitric acid production

Since nitric acid is used in fertilizer production a reduction in fertilizer production will lead to a reduction in nitrous oxide emissions from nitric acid production.

Human sewage

Fecal matter, output from garbage disposals and washing machine effluent are all rich in nitrogen and the microbial activities in our sewers produces 13 million metric tons carbon dioxide equivalents of nitrous oxide a year that is released into the environment.

An Introduction to Energy

Solutions to Energy Problems

All living things large and small require energy. Those who use it more efficiently prosper.

___Switching from the Heat Engine

Although there have been some improvements to the heat engine (hybrid vehicles) we are faced with the fact that it is inherently inefficient.

The best solution to the inefficiency problem of the heat engine is not to use it in the first place.

The following sections will cover each of these technologies along with the challenges they pose to their implementation and use.

A persistent problem with the following technologies is their cost of implementation which is considerably more expensive than fossil fuel-based technologies that exist today despite the fact that costs of operation are considerably less and payback on initial investment is high compared to the heat engine.

An Introduction to Energy

_____Hydrogen Fuel Cell

One of the major benefits of a hydrogen fuel cell machine is that it emits water as a waste product instead of carbon dioxide or monoxide gas.

The hydrogen fuel cell idea was born during the space race to land a man on the moon. It was a way to provide two essential resources to the astronauts in space namely drinking water and electrical power.

The way the hydrogen fuel cell works is that hydrogen and air are mixed and converted to water using a catalyst. The reaction produces an electrical current that can then be used to operate an electric motor or charge a battery.

But as with many new inventions the rush to develop a vehicle that uses hydrogen fuel cell technology glossed over the fact that it takes energy (usually fossil fuels) to generate the hydrogen. And when you realize that hydrogen is separated from water by using electricity the full problem presents itself.

First we must generate electricity then we need to use the electricity to make hydrogen then the hydrogen is used to make electricity again to run the electric motors and according to principle 2 that was discussed earlier that means that we are putting more energy into the process than we are getting out and the emission of more greenhouse gasses. Clearly this is not a viable solution.

_____Nuclear Power

The initial numbers looked promising a while ago when the nuclear fever was at its highest. One pound of uranium costs $50, is equivalent to a million gallons of gasoline and it does not produce one ounce of carbon dioxide.

A nuclear power plant has three parts. It has a nuclear reactor chamber filled with water in which nuclear fuel is placed in the form of rods that are separated by spaces. In the spaces are removable control rods that can move to increase and decrease the nuclear reaction between the fuel rods. As the fuel rods react with one another heat is generated that in turn heats the water converting it to steam.

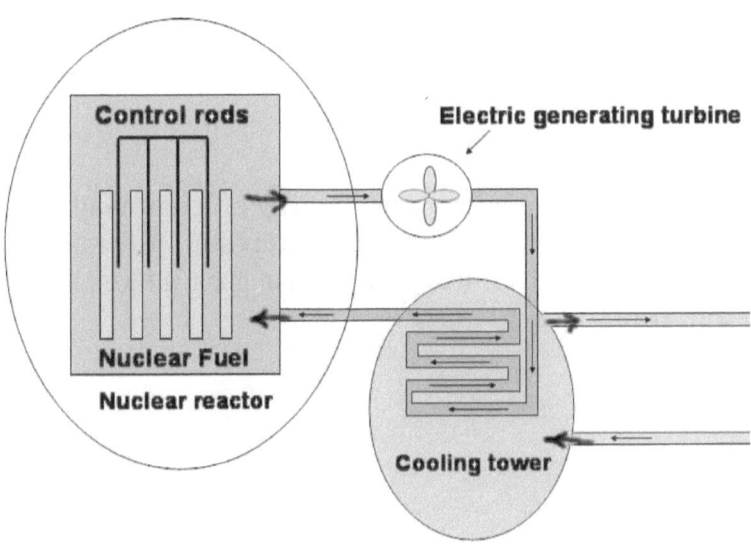

Graph showing a schematic of a nuclear power plant

The steam travels to the electric generating engine then travels to the cooling tower where it is cooled then returned to the nuclear reaction chamber as water.

Unfortunately, design, construction, maintenance and disposal of the radioactive waist had increased the cost of building nuclear reactors reducing their efficiency significantly. Coupled with the fact that nuclear power plants generate electricity using the old and inefficient heat engine technology most of the fuel that is consumed and the waste that it generates resulted in no benefit. Today the cost of nuclear energy is at par with coal fired plants. In the eyes of many, it is a viable solution to energy efficiency despite the fact that it has many other side effects such as safety and radioactive waste products that are dangerous to the environment in other ways than the green house effects of carbon dioxide.

_____Solar Power

Solar power generation is the most energy efficient way to produce power therefore it is of great interest to us besides it does not produce any harmful greenhouse gasses like carbon dioxide and smog.

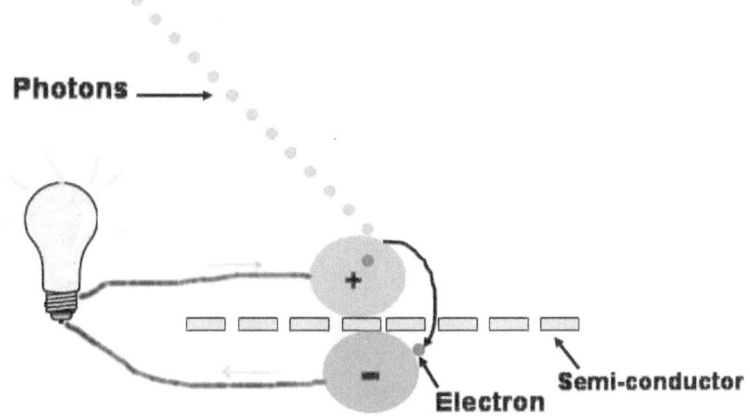

Graph demonstrating how a semi-conductor works

Photovoltaic (PV) devices use light to generate an electrical current. Some materials exhibit a property known as the "photoelectric effect" in which they absorb photons of light and release electrons. When these electrons are captured, an electric current is generated that can be used to do work.

A photovoltaic cell, also called a solar cell, is made of a semiconductor type material, such as silicon. The term semiconductor means that the material allows the passage of electrons through it in one direction only.

A thin semiconductor wafer is specially treated with a material exhibiting photoelectric properties on the side that will be exposed to the sun. This side of the wafer is also known as the electron donator or positive side of the photo voltaic cell.

On the back side of the wafer (the side that is not exposed to the sun) the photo voltaic cell is coated with another material called the electron receptor, or negative side.

When exposed to sunlight, electrons are knocked loose from the atoms in the donator side and travel to the receptor side. If electrical conductors are attached to the donator and receptor sides the electrons create an electrical current as they move back from the electron receptor side to the electron donator side. This current can be used to do work such as power a light bulb or charge a battery.

Photo voltaic cells have an optimum efficiency of around 5-15% in converting light energy to electrical energy. This is because only a narrow band of the light spectrum (specific wave lengths) can be used to excite electrons out of place in the electron donator side, and different photoelectric materials are excited by different wave lengths.

Optimizing sunlight capture can be done by including concentrators which are like mirrors that gather light from a wider area and reflect it onto the photo voltaic cells. In addition, the surface coating of a cell is important in decreasing the number of photons that bounce off of if thus decreasing its ability to perform.

Photo voltaic cells only operate during the day light hours. This poses a problem when their electric output is needed during the night. To solve this problem, they are often used to charge batteries during the day in stand-alone systems. However, batteries are not 100% efficient further reducing their efficiency. PV systems can also be connected to an electric utility's power grid where excess power is sold to the utility and later purchased back at peek demand periods.

Photo voltaic cell output is also affected by cloud cover because clouds reduce the number of photons that hit the cell's photoelectric side further reducing its efficiency.

The photo voltaic cell needs the sun to be perfectly perpendicular to the photoelectric side to achieve its maximum potential efficiency but unfortunately the sun moves across the sky at different angles. Additional hardware in the form of sun tracking mounts can be installed to keep the sun perpendicular to the PV cell.

The environment also poses performance problems such as pollution and dust that decreases the number of photons hitting the PV surface. Sand can also etch away at the surface of the PV cell affecting its photoelectric properties and or the properties of any other coatings designed to improve performance.

Basic costs, beginning with a photo voltaic system that produces 600 watts and operates several lights, a TV, stereo, microwave and water pump (not at the same time) costs about $8,000.

More precisely, the cost of a home system is around $4.00 per peek watt. A peek watt is a standard way the industry measures the performance of solar cells and is done at the factory by shining a 1000-watt source of light per meter squared on the solar cell and measuring its output. The cost of the solar cells themselves is around 40% to 50% of the total installation cost.

A home in Sacramento, California can expect to generate about 20%-30% of its electrical needs from a photo voltaic system of medium size.

An Introduction to Energy

Another way to improve the efficiency of our economies is by harnessing the power of the wind and using wind turbines to generate electricity.

Wind systems can be stand-alone or be connected to an electrical grid run by the electric utility company. Unlike solar energy they are available to generate electricity 24 hours per day depending on geographic wind patterns. They are mechanical, requiring periodic preventative maintenance to improve their efficiency and life expectancy. Increasing size and scale of the generating system can reduce costs and improve efficiency substantially.

A wind power generator consists of a propeller turning an electrical generator, mounted high off the ground. There are two popular ways to capture wind energy, both having to do with the shape of the propeller blades. Broad blades are used to generate low speed torque, and are well-know from the windmills in Holland or the old water pumps used on farms. Electric generating propellers are much slimmer and designed for higher speeds that are required by generators.

Electric generators have an optimum turning speeds at which they generate electricity efficiently. Many designs require wind speeds of around 30 miles per hour. Smart generators stop turning at less or greater than optimum speeds to save wear and tear on mechanical parts. Other smart generators have variable-width propeller blades and transmissions to maintain optimum generator speeds within a wider range of wind speeds. The theoretical maximum efficiency rate for a wind system is 51%.

The total installation cost can be expressed as a function of the wind system's rated electrical capacity. A grid-connected residential-scale system (1-10 kW) generally costs between $2,400 and $3,000 per installed kilowatt. That's $24,000-$30,000 for a 10 kW system.

A moderate-scale (10-100 kW) is more cost-effective, costing between $1,500 and $2,500 per kilowatt. Large-scale systems greater than 100 kW cost in the range of $1,000 to $2,000 per kilowatt, with the lowest costs achieved when multiple units are installed at one location. In general, cost rates decrease as system capacity increases.

Operating costs include maintenance and service, insurance, and applicable taxes (or tax credits.) A rule of thumb estimate for annual operating expenses is 2% to 3% of the initial system cost $.01 to $.02 per kWh of output.

Despite these initial and ongoing costs (which are considerably greater than an internal combustion engine based generator) a wind turbine will pay for itself within a matter of a few years.

_____Wave Power

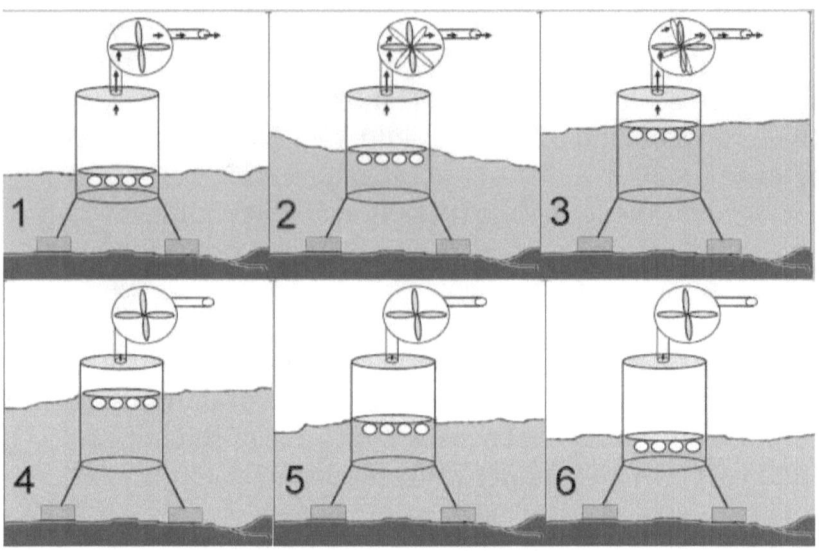

Graphic frames showing a wave coming through and affecting the operation of a wave power machine

The least developed source of efficient electric power is wave power. There are a few of them around mostly in experimental stages.

Wave power is generated from waves in the oceans. The simplest consists of affixing a tube half way under water that has a floating diaphragm. The tube is closed at the top except for a valve that allows the passage of air out and another that allows the passage of air in. The valve that allows air out in turn drives a turbine electric generator.

In the gully between waves the diaphragm moves lower filling the tube with air from the intake valve. As the ebb of the wave approaches it raises the diaphragm up pushing the air in the tube through the outlet valve and into the electric generator.

There are no firm numbers on the costs of any particular type of wave electric generator and most cost about the same as a windmill electric generator. It does have unique problems such as salt water corrosion, interference with marine navigation and the fishing industry. But the return on investment is in the same area of magnitude as wind generators.

___What do we do Till Then?

Until nuclear, solar, wind and/or wave power become mainstream there are ways to increase efficiency of energy consumption. They include better city planning, and changes in our everyday habits.

There are several opportunities available to city planners that can reduce carbon dioxide emissions that include business to residential zoning, mass transit and planning on smaller scales.

Finally, each of us can make a difference in the way we use the heat engine through a few changes in decision making and habits.

_____Dwellings

Underground Dwellings

The human body is an adaptive machine. It is able to tolerate and change to accommodate many environmental conditions, including temperature. The ideal temperature for the human body is around 70 to 80 degrees. This allows it to function and get rid of excess heat generated through physical and mental processes. The temperatures that the human body can adjust to range from about 50 to 90 degrees. At 50 degrees, moderate physical activity or an additional layer of clothing can restore considerable comfort. At 90 degrees, the body's natural cooling system, sweat, will also increase its ability to rid itself of excess heat.

One option is to relocate if the weather in your current place of residence is too hot or cold. However, this may not be an option.

If you can't move, then consider moving underground. It is well known that the deeper you go the more stable and comfortable the temperature is. For example, in West Texas a study was made of soil temperatures at various depths. Although the temperature varied from 55 to 95 degrees from winter to summer respectively at the surface, the temperature ranged from a most comfortable 68 to 77 degrees at ten feet down. This eliminates the need for heating in the winter and cooling in summer. This idea is not new and is practiced throughout the world and by Native Americans.

An underground dwelling in Arkansas

The following graphic is of a conceptual drawing of a structure that does not require heating or cooling in winter or summer where the outside temperature ranges from 50 to 95 degrees. This underground structure is built in a mound of dirt to avoid high ground water levels or flash flooding. It is constructed in much the same fashion as an artificial botanical pond.

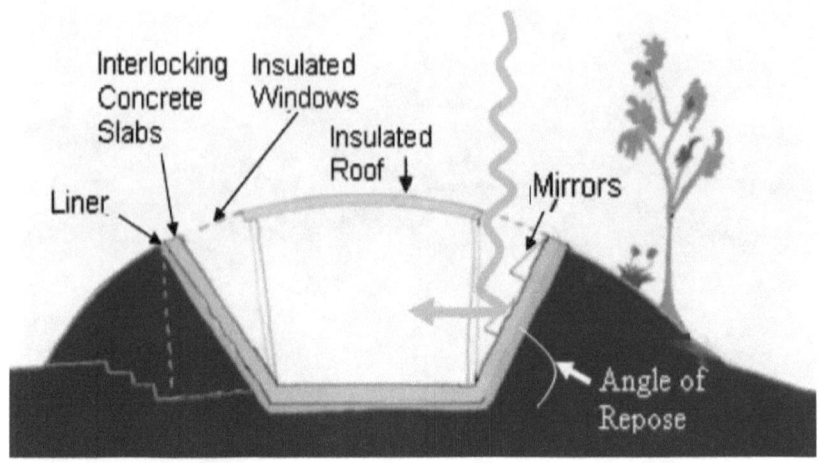

Conceptual drawing of a modern underground house

First, a rubber liner is placed in the dugout area and lined with concrete slabs at the angle of repose. The living area is then built within this cavity with a well-insulated roof. The gap between the conventional square structure inside and the slanted walls is covered with insulated rain-proof windows. These windows allow much-needed light, and in this particular structure, the light from the sky is reflected into the living cavity by the use of mirrors. Many variations on this basic theme are possible.

The Shape of Things

One observation from nature is that reducing the surface area exposed to the outside elements conserves the effort of keeping its interior cool or warm. This ideal conceptual form is the sphere.

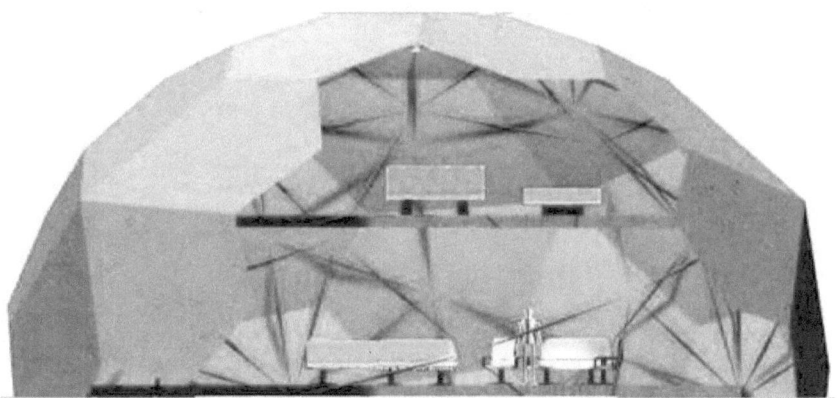

A Geodesic house

Some innovations have come from this, one of which is the geodesic dome structure. In the graphic above is a conceptual drawing of a geodesic dome living structure consisting of two floors.

These structures became very popular in the 70s and 80s and were cost effective to make and build. Some come in kits that one or two people can put together. Besides conserving energy, they are strong and durable, withstanding storms that other structures can't. Today, kits can be acquired made of foam for simple installation and supper insulation.

But say that you aren't comfortable with an underground or dome-shaped house. You can use some of the ideas these structures bring to the energy conservation arena, like building a house as close to a cube shape as possible.

The following are conceptual drawings of one and two level structures all with 5,000 square feet of living space. Please note the reduction in surface area.

A one-story structure exhibiting a surface area of 13,000 sq. ft. and a two-story structure with a surface area of 9,000 sq. ft.

Please note that a three-story structure will have a surface area of 6,800 sq. ft.

As can be seen from the above drawings, energy efficiency can be improved by 45%. For example, if energy costs are around $200 a month during the winter, moving from a 5,000 sq. ft. one-story house to one that is the same square footage but is three levels could reduce energy costs to about $110.

There are other benefits to building houses in the shape of a cube in that decreased surface area translates into less insulation to be installed, as will be demonstrated in coming chapters.

The Sun

The sun is an important source of energy in the form of heat. In the north it is welcome anytime, but in the south, it can be a burden when it comes to home comfort. The following discussion will start with over-exposure to the sun and what we can do about it. At the end of the chapter, we will talk about the welcome sun in the north and how we can capture its benefits.

Colors

Walls exposed to the sun can be treated to reduce or increase the absorption of the sun's energy. In the hot south, the desire to prevent heat from building up in our houses is opposite to those in the cold north, where the sun's rays are a welcome sight.

Generally, light colors reflect heat, while darker colors absorb heat. So it may be wise to paint your sun-facing outside walls white, light blue or light green to keep the walls cool and prevent the migration of heat from the walls to the interior of the house in hot climates. In chillier climates choose dark colors to squeeze every calorie out of these precious rays.

The "do-it-your-self-ers" building a house can do some experimentation with color by letting a box sit outside for an hour or two with a thermometer inside. Note how hot it gets with different paint colors you intend to use. You'll be surprised at the effect of paint color have on the temperature inside the box.

A better experiment would use a box constructed of the actual materials used in your final product (drywall, insulation, siding, etc.)

If you have a dark-colored house, an air conditioner, and you decide to paint your house a lighter color (say, yellow), you can expect to have a return in calorie investment of around 1500% every month in the summer! (Assuming it took you 4 gallons of paint to cover 50 feet of exterior sun-exposed walls of a 1500 ft^2 R17 insulated house, and that it reduced your interior house temperature by about three degrees.)

Windows

Windows are transparent and allow sunlight directly into the house. This can have good or bad consequences, depending on where you live.

A load bearing wall may have no windows, other walls can have 50% of the surface area in windows. In hot climates we need to prevent this energy from entering the house. The simplest way to do this is to place curtains on the outside of the window. It is a misconception that we need to place the curtains on the inside: energy becomes trapped in the house. Outside curtains trap the energy outside and it is simply diffused into the outside air.

In colder climates, placing dark curtains inside will trap the sun's energy in the form of heat inside the house. You don't have to limit your efforts to trap the sun's energy to curtains. Painting the interior of your house in dark colors will have a similar effect, and allow you to enjoy the sunlight as well.

You can purchase insulated windows that can prevent the migration of heat from the outside and vise versa. Despite the short life expectancy of these windows (about five to ten years) they are well worth their high cost.

A comparison between aluminum and well insulated widows for a 1,500 ft^2 house located in the temperate northwest demonstrates that in the first year, a conservation of about 3 billion calories is possible, resulting in substantial savings in heating bills.

Insulation

When the temperature outside is 20 degrees and all your doors and windows are closed, chances are, your house will become the same temperature through conduction of heat from the interior to exterior.

The best way to handle this problem is by the use of insulation, which hinders the conduction of heat. This conserves heat so you run the furnace or air conditioner less often.

Insulation comes in several forms, but we will talk about fiberglass here simply because it is one of the most popular materials. It is rated by "R values" which reflect its ability to resist the conduction of energy, the higher the "R" rating, the greater the insulation's ability.

The drawback to insulation is that it seals the inside quite tightly and may cause ventilation problems. However, there are ways to ventilate the inside of the house without opening doors or windows and defeating the purpose of insulation. A popular way to ventilate and conserve heat or cold is to use a heat exchanger. Heat exchangers have been use for a long time in industry but haven't been wide-spread in homes because of their higher cost in comparison to the price of oil.

High performance heat exchangers can deliver 90% or more heat or cold conservation.

Heat Exchangers

The principle of a heat exchanger was found in husky dogs in Alaska by biologists. They discovered that arteries in the legs of huskies run close and opposite to the veins, to conserve heat around their vital organs.

A simple heat exchanger consists of two pipes directing air or water in opposite directions. The neat thing about them is that the temperatures of air or water are pretty close at both ends if the pipes are long enough. The length of the pipes allows the hot and cold medium to be coupled long enough to allow the heat to be transferred from one to the other.

There are many heat exchange designs depending on the application to which they are to be applied. You can purchase heat exchangers that can conserve anywhere from 50% to 90% of the heat.

Roof

Luckily the roof works like a buffer to heat if the house is designed properly. Originally the roof was designed to protect the structure from rain. In doing so it creates a cavity over the house. These cavities are supposed to be well-ventilated to prevent problems of moisture buildup and resultant complications such as mold and rot. Unfortunately, the degree of ventilation for moisture prevention doesn't solve problems with heat. It is common for the attic to reach temperatures 150 degrees which can make living under it unbearable. A simple solution to this problem is to increase the number of roof vents to let the heat escape. Along with the heat escape vents, you must have sufficient air inlet vents at the base of the roof to allow cooler air to replace the heat.

There is an array of vents available ranging from $10 to $150. The more expensive ones have fans that can be turned on and off depending on the temperature and or humidity of the attic. Some even have fans that operate on electricity from solar panels regulated by thermostats.

Exposure

The problem with the sun in the south is that it is too intense especially when it hits our homes directly. Most houses have five sides above ground.

One way to reduce exposure to the sun is by building homes with the least exposure to it, with the narrow end of the house facing the sun.

Reducing the direct exposure of the house to the sun can result in a 50% to 90% decrease in the amount of energy (heat) that it is exposed to, depending on the surrounding terrain.

In the north the opposite is true and setting up the house so that the sun hits the maximum area of the exterior walls is the rule.

Shade

To further decrease the sun's power we can plant trees or shrubs between the house and the sun, so that only indirect sunlight reaches the house, as can be seen in the above drawing.

Evergreen trees are a good choice if the winter months are also hot. To pick a suitable tree for such a project, you need to know the angle of the sun from the horizon; this will dictate how tall a tree to plant and how far from the house it should be planted.

The above illustration shows how to estimate the height of a tree from your house to provide maximum protection from the sun's direct rays for one, two and three story houses.

Please note that the protractor is placed at the roof line and that the tree is placed well away from the structure. The farther out the tree the taller the tree should be. Caution should be taken when placing a tree close to a structure because of damage it may cause due to mold, etc. Always consult your local government building codes.

The reason why you should estimate the height of the tree from the roof of the house is to make sure that the whole side is shaded. Otherwise, the top floors may not get the full benefit.

Another method of reducing the exposure to the sun is to building a porch, arbor or trellis near the house. They can support vine plants like ivy and grape, as illustrated above.

You can create indirect sunlight with above window shades on the sunward side of the house.

When planning the use of vegetation to create shade, you must also take into consideration the effects it might have on air circulation and make sure that there is enough air movement to prevent the establishment of mold and mildew.

As we move north winters are colder, consider planting deciduous plants (plants that lose their leaves in the fall) to protect the house from the hot sun in the summer and increase exposure to it in the winter, to help with heating bills. Instead of pine trees, plant maple, and instead of ivy, plant grapes.

In addition, shade structures can be built to protect the house from the sun in the summer and expose it to the sun in winter. A porch or other shading device can cover the exterior walls of the house in the summer only. As the sun is lower on the horizon in the winter months, rays can sneak under the shade structure to warm the walls of the house.

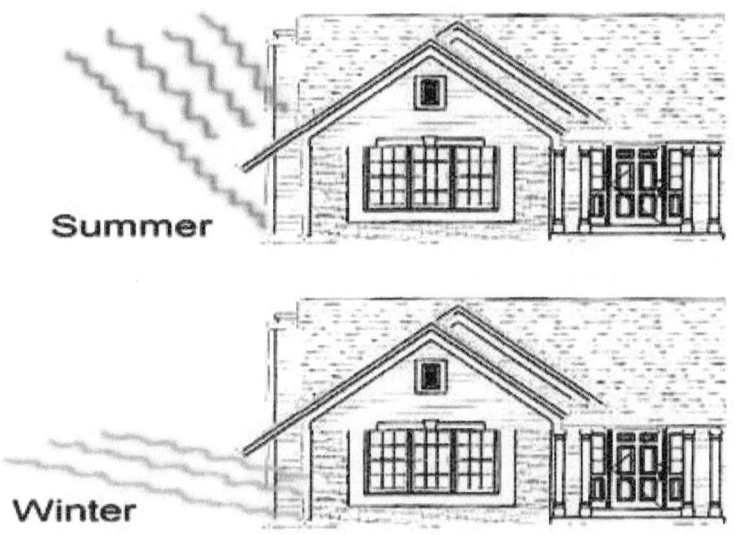

Summer

Winter

Far north, it may be that you do nothing and allow the full impact of whatever sun you have to reach your house.

In planning the position of a tree on your property, it is also beneficial to consider its effect on wind and breeze. In hot climates, it would be advantageous to have trees that allow the cooling effect of breezes. For single-story houses in hot climates, consider planting trees with few low branches.

In cool or cold climates breeze and wind may increase your cost of heating due to the wind chill factor. In this case, consider planting trees with low-lying branches like evergreens or shrubs that break the effects of the wind.

An Introduction to Energy

_____Geothermal Energy

Geothermal energy is found in the crust of the Earth and although the crust is miles and miles thick there are cracks and bulges in it that allows the extremely hot magma at the center to seep and exist closer to where we can access it.

The hot magma is like having a source of heat without having to burn anything and produce greenhouse gasses. With the heat from this magma we can power heat engines that can in turn generate electricity. It's like having a nuclear power plant without the nuclear fuel.

The idea behind geothermal energy is to secure two pipes that reach the same source of magma and injecting water down one and allowing the generated steam to rise through the second one where it is captured to power a steam engine like a steam turbine.

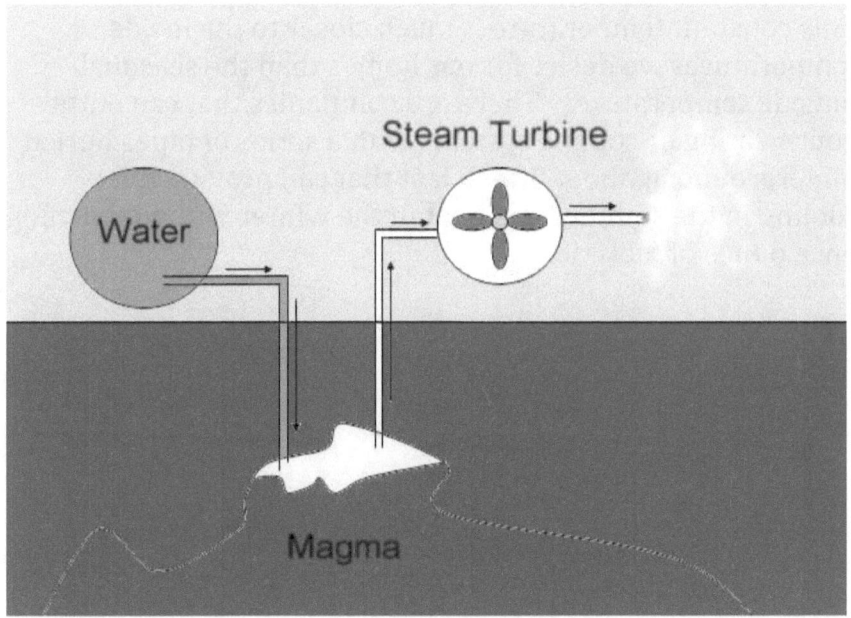

Graph showing the different parts and their relationships of a geothermal plant

Once the technical problems that exist for this technology are solved, it can generate only supplemental energy to other sources of energy that exist today. These technical difficulties include corrosion and mineral depositions on the equipment that require constant repair, very deep drilling of the magnitude of miles in the crust of the earth and the fact that the crust of the Earth does not have enough accessible points to generate all the electrical or heating power needs of the Earth's population.

Fortunately, there are other ways to access some of the warmth of the earth to supplement our energy needs. If you dig about ten or fifteen feet underground you will find that the temperature there remains pretty constant from winter to summer as can be seen from the following graph from Texas.

This constant temperature is much closer to the inside temperatures we desire for our homes than the seasonal outside temperatures. There are companies that can outfit your heating or cooling systems with a series of pipes buried underground at about fifteen feet that can provide this cooling in the summer and heat in the winter without burning once ounce of fuel.

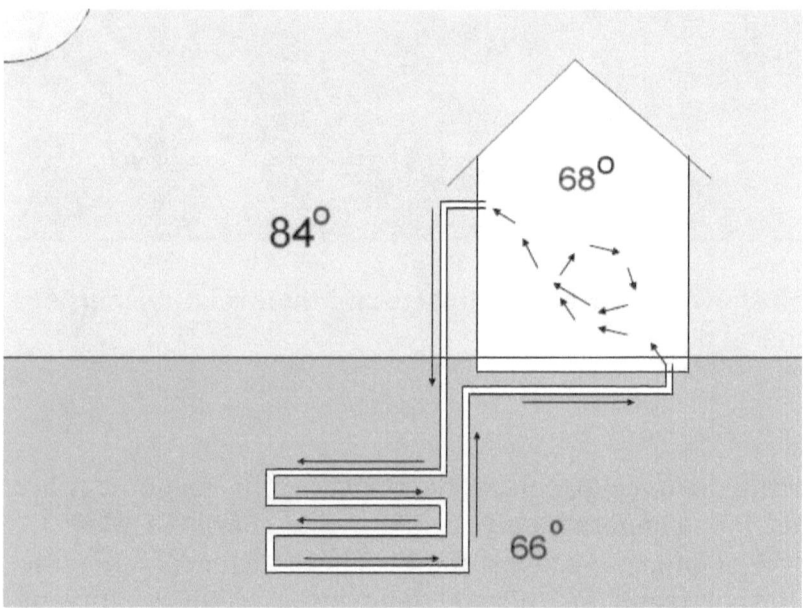

Graph showing how a house can be heated in winter and cooled in summer using deep soil temperature sinks

_____City Planning

City planning is a community effort and is sensitive to the wants and needs of its people. In large cities smog was the primary problem these communities faced and to deal with it they implemented mass transit among other initiatives. However the major reason was not per say to reduce pollution. It just happened that implementation of mass transit did reduce energy consumption and thus increased efficiency and lowered costs. Even small cities are working on solving this problem. For example, the city of Portland, Oregon has setup an organization to oversee the efforts to convert city services to use "sustainable" resources and set the lead for the community.

Mass transit:

Although there are many arguments against mass transit which include the involvement of large private and public subsidies and that it does not provide direct pickup from point of origin to the destination point the evidence points that it saves oil consumption and thus reduces carbon dioxide emissions as well.

For example, a bus weighs 10,000 pounds while the average car weighs 4,000 pounds and using quick math it can be seen that if a bus eliminates 3 cars (12,000 pounds) from driving around town it is worth the switch from private to mass transit. And each additional passenger reduces fuel consumption by the equivalent of moving a two ton vehicle (even more if SUVs are factored into the equation.) It is true that rider ship is low (around 30 to 40 percent of available seats compared to other developing nations) but the estimates also indicate that this low rider ship reduces fuel consumption by 50% meaning that for every gallon of fuel consumed by the bus two gallons are saved; a net reduction of 27 pounds of carbon dioxide emissions.

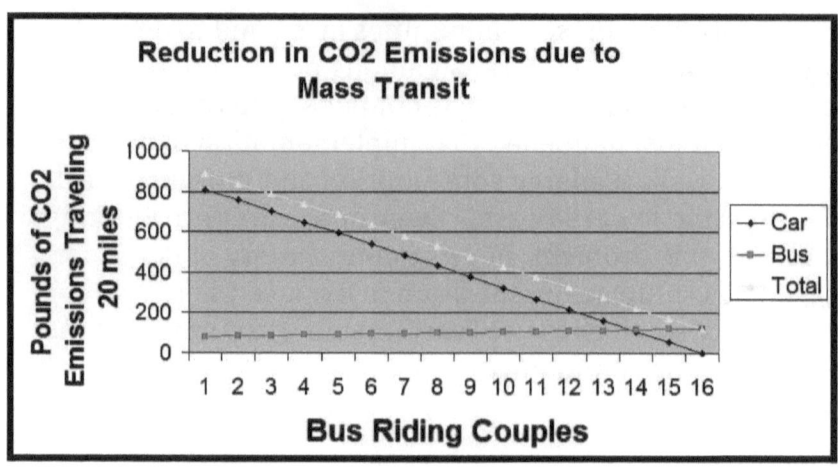

Graph showing the results carbon dioxide emissions and thus improvement in energy efficiency when mass- transit is introduced into a community

The above chart measures carbon dioxide emissions of 16 couples as they shift one couple at a time from driving their own cars to riding the bus.

The adoption of mass transit is in its infancy in most cities in the US and there are definite hardships. For example, mass transit riders need to walk to bus stations and walk from bus stations to their final destinations. As it is today cities are not designed well for pedestrian traffic but for private transportation. Major redesigns of the inner city need to be made to make them pedestrian friendly and encourage bus rider ship. An example of three decades of planning for pedestrians can be seen in the city of Curitiba, Brazil.

Decentralization

Another method to increase energy efficiency is through the elimination of Mega factories and Mega retail outlets by encouraging smaller scale development so that people live close to their places of work and shopping centers. This would reduce the distances that we would have to travel each day to get to and back from work or the grocery store. The following graph is an illustration of the effects of eliminating large centralized grocery stores and replacing them with many smaller ones near neighborhoods and the effect it would have on fuel consumption.

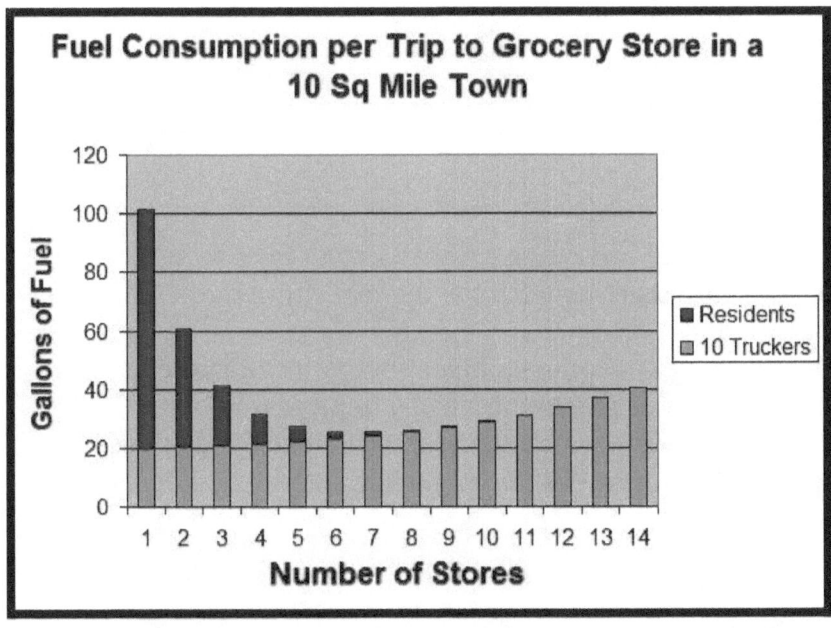

Graph showing the effects on carbon dioxide emissions when grocery stores are decentralized

An Introduction to Energy

The above graph is a little complicated but if you follow the bars from left to right and the number of pounds produced by residents (lower bars) and grocery trucks (darker upper bars) dip dramatically around 8 stores every 10 miles for a city of roughly 100 square miles in area. This is roughly a store every mile or so within the city. This dramatization indicates that the concept of Mega-stores that service large city areas causes a dramatic increase in fuel consumption.

Similar planning for factories and other places of work can have a similar and significant effect on fossil fuel consumption to say nothing of the reduction in traffic density.

_____The Auto

Each individual can make a difference in the amount of fuel consumption starting with the biggest contributor; the auto.

The following are some changes that you can make in your auto purchase or use habits. They depend on your willingness to use manual power, scale back your purchases to better fit your needs and make efficiency wise decisions.

1- If the distance you need to go is a mile or two, walk it;

2- If the distance you need to go is two or three miles then ride a bicycle;

3- If you are single, look into buying a motorcycle;

4- When shopping for a vehicle:

a. Choose a vehicle that fits you like a glove. This means avoid buying one to accommodate casual uses. For example, if you need a four seat car and once in a while it would be nice to accommodate six people then buy a four seat car and rent a limousine on special occasions. The same goes for a four-five seat pickup; you can always rent a pickup when the time comes;

b. Look at the EPA mileage sticker very closely and select the one with the most miles per gallon from your choices;

c. If EPA mileage stickers are the same between two choices, chose the lighter car;

d. Given EPA mileage stickers that are the same between two choices, chose the one with the smaller engine by looking at the engine displacement;

e. Given If EPA mileage stickers that are the same select the car with the smaller and aerodynamic body;

f. Electric or hybrid cars are more fuel efficient.

5- If you have a car that doesn't have an RPM gage, install one and try to avoid going over 3,000 RPM while driving;

6- Make as few trips per day as possible by grouping as many of them as you are able to in one outing by taking out a few minutes for planning;

7- Use routs with the fewest stops;

8- Choose the shortest routs;

9- Keep track of the miles you travel;

10- Keep your tires properly inflated;

11- Keep your vehicle in the best working order;

_____Recycling

Like Mother Nature we have learned to do more with less energy by recycling; by making cycles of our resources; by using and reusing what is available to us.

To make one ton of glass we must collect 1,330 pounds of sand from the desert or the sea shore and clean it. We also need to mine for 151 pounds of feldspar and 433 pounds of limestone. Then we need to provide 433 pounds of soda ash. All need to be transported to a glass factory where they are mixed and heated up using fossil fuels and the emission of 28 pounds of carbon dioxide. to make one tone of glass.

That is pretty much what is involved in producing our consumables from raw materials; a lot of work and energy. If we recycle the glass that is already in use we can reduce emissions by 9 pounds of carbon dioxide per ton and if we recycle most of the glass in the US we can reduce emissions by 190 million tons of carbon dioxide a year.

In a similar fashion, the raw materials needed to make all the paper that is consumed each year in the US are 250 million trees. Fortunately, we are currently recycling about half the paper we use in newspapers, for example, and so before us is an opportunity to save another 125 million trees and 313 million tons of carbon dioxide emissions a year. In addition, the trees that we don't cut down or the trees we can plant will act as carbon sinks for the next hundred years or so.

Steel is much like glass in its requirements for mining and heating the raw materials to high temperatures. And although almost all autos are recycled these days, "tin" cans are not all recycled and if we recycle them we can reduce carbon dioxide emissions by 75 million tons a year.

Then there are the billions of plastic bottles that are used for all the beverages we drink. By recycling them we can reduce emissions of carbon dioxide by another 72 million tons a year.

It shouldn't be a surprise that by just recycling the above we have reduced emissions of carbon dioxide by half a billion tons which is about 8% of the total emissions in the US. Other consumables that we can recycle are:

> Mixed paper and cardboard

> Scrap aluminum

> Motor oil

> Auto batteries

> Printer cartridges

> Computers

An Introduction to Energy

Human kind has been using fuels from biological origins ever since we discovered fire. The types of fuels used then (and still used now around the world) are wood, straw and dung or just about any waste material that can burn.

Today, bio-fuels is a term used to describe the process of making a more convenient fuel out of those waste materials, usually in liquid form, to accommodate its used in the various heat engines that exist today.

The driving factors behind bio-fuels are many. They include but not limited to they may lead to energy self-sufficiency, help the conservation recycling efforts and reduce the production of greenhouse gasses.

The way that bio-fuels reduce greenhouse gasses is by providing a fuel that is part of the plant growing cycle and as a result zero net production of carbon dioxide occurs. In other words a kernel of corn is converted to a fuel and the emissions that result from burning it in a heat engine go back into growing another kernel the next year.

If we take a stalk of corn, the easiest part of it to convert to bio-fuels is the corn kernels. That is because it consists mainly of sugars and starches. The stalks, leaves and husks are mostly made of cellulose which is harder to convert to bio-fuels because they don't break down readily.

There are many researchers working on how to breakdown and make bio-fuels from cellulose which is also found in wood. For example, some are studying the chemical reactions in the guts of termites which can breakdown wood fiber into starches and sugars which in turn can be converted to bio-fuels.

But in 2007 the price of milk jumped up by fifty cents a gallon as a result of a rise in demand for corn caused by milk farmers and bio-fuel manufacturers competing for the same resource. This brings us to the reality of the impracticality today of bio-fuels. If the US is to achieve true energy independence every inch of arable land will have to be planted with bio-fuel crops.

That fact has turned the attention of some scientists and developers to the seas and oceans. They are researching ways of converting seaweeds into bio-fuels. But again, we must ask the question, "How will that impact the ecology of fisheries and more generally the oceans themselves?"

_____Heating Water

In the south, you can almost do without a water heater by using a solar water heater and save your usual utility energy costs.

A solar water heater consists of a solar collector and a water tank. The collector is made of an insulated window box with coils of copper pipe on a copper sheet supported by a strong flat support which are painted black to maximize the absorption of heat from sunlight. As the water in the copper pipes heats, it moves to the insulated water tank, providing a source of hot water to the house night and day.

As you move north, the angle of the sun and the presence of clouds reduce the amount of heat that can be absorbed and thus the availability of hot water, especially in the winter. A solution to this problem is to couple a solar hot water system with an ordinary water heater using a heat exchanger so that the later can warm the solar heated water up your desired temperature, while still saving you significant water heating costs.

There are many designs to solar water heaters, from simple plastic units that hang on a roof, to more elaborate ones consisting of copper or brass tanks hidden in the roof, with the solar collector fastened on top.

The key to any solar collection device is its placement with regards to the sun. The collector should face the sun which can be tricky because the sun moves across the sky in an arc. There are many factors dictating the size of the solar collector: the angle of the sun, the presence of clouds, smog, and the length of daylight, the number of people in the household, and the uses of hot water. It is suggested that you find a household similar to yours that uses a solar water heater and get your ideas from them. Consult a solar hot water vendor. You can also purchase solar panels arranged in series or in parallel and start with the minimum you think you might need and then add on later as your particular demands become clear.

Hassan Rasheed

www.ingramcontent.com/pod-product-compliance
Lightning Source LLC
Chambersburg PA
CBHW032018170526
45157CB00002B/748